Television Graphics
—from pencil to pixel

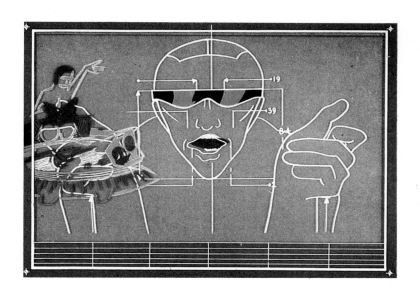

Douglas Merritt

Television Graphics

–from pencil to pixel

Trefoil, London

First published 1987
by Trefoil Publications Ltd
7 Royal Parade, Dawes Road, London SW6

British Library Cataloguing in Publication Data
Merritt, Douglas
   Television graphics : from pencil to
   pixel.—(Trefoil design library).
   1. Television graphics
   I. Title
   741.6   PN1992.8.G7

   ISBN 0-86294-070-2

Designed by Douglas Merritt.
Set in Futura Light by Suripace Ltd, Milton Keynes
Printed in Spain by Novograph SA, Madrid

Contents

Foreword

'And what is the use of a book', thought Alice, 'without pictures and conversation?'*

And what is the use of a book on television graphics where there can be no moving pictures or sound?

In spite of the limitations of reproducing television animations and stills from video this book attempts to record what graphic designers have done, and to explain some of the background to a subject that has become increasingly complex in the past few years. It is not a 'How to do it book', rather 'How it is sometimes done'.

Does it now appear that television graphic designers use computers for everything they do – even to sharpen their rarely used pencils? Is their contribution to the vast audiences so enveloped in technology that the jargon conceals what they do?

The hope is that graphic design output explains and entertains in transmission – using improved methods and new equipment that was not available until fairly recently. Viewers should never be aware of the problems: Film versus video production: cel animation versus computer animation.

Only in the context of presenting programmes can graphic design have any true value. It must never seek to be an end in itself.

Although the 60s produced a few books on the emerging profession, at about the time it became of age, the amount of printed information has been very limited. Grateful reference is made to the authors of these books throughout my text. Now, with their help, the first half century can be reviewed. For those, who like Alice, are prepared to be curious, a brief view can be taken 'Behind the screens'.

The progress and the changes in graphic design for television are directly related to the development of television transmission and engineering standards and it is important to recall the technical improvements which have taken place in the past fifty years.

All of these have controlled the style and general character of graphic presentation; from the crude monochrome picture on the very small screen of the early days, (the words 'the small screen' were coined to portray the whole new medium), to the current experiments in full-colour high-definition television – known as HDTV.

50 years of television

Many people are surprised to learn that half a century has already passed since a regular television service was first available – albeit to only a handful of viewers – in the United Kingdom. The Royal Television Society, even more surprisingly, reached sixty years in 1987.

The BBC first transmitted pictures from the Alexandra Palace on 2 November 1936 and has continued to provide a public service since then with the exception of the close-down throughout the Second World War from 1939 to 1946.

The social influence and power of television can appear to exaggerate the importance of the work of those involved in its production. A series of articles in *The Observer* at the beginning of 1987 produced a survey showing 95% of the British population spend an average of 25 hours a week watching television. That is nearly a quarter of their waking hours every week!

World-wide television

At present 134 countries broadcast television and the way they chose to organise those services is reflected, or conditioned by, their social and political systems. A study of the variations and the complexity of these would be a vast undertaking, but all of them now recognise and use the services and skills of graphic designers. This was not always the case and only gradually has graphic design begun to play a more important part in what we see on our screens.

Euro-television

Most European countries have a mixture of state or government control over their television stations and these are funded by

taxation, or licence fees, while running along side them are the commercial stations where programme-making is supported by selling advertising or by sponsorship. The progress in all programme-making depends on the size of the audience and the revenue available and graphic design has grown in relation to these two factors.

Competition for audiences between the two systems, public and commercial, has done much to keep standards high in all areas of programme-making. The first graphic designers arrived at the BBC just before the first British commercial stations went on air.

The French government is about to privatise one of the two state systems, TF1, while retaining the more profitable service Antenne 2. The staff and the public are concerned by this because the control of the programme content is fundamental to the work and the output of the station.

Whatever the system, or the form of the organisation, the type of programmes which can be made will have a direct influence on the style, quality and range of graphic design. By now the function of graphic presentation is such an integral part of the programmes it can no longer be considered an optional extra.

American TV

The American experience of the growth of television has been, as might be expected from other aspects of her history, without the constraints of many other countries.*

Leesburg, a small town in Virginia, was once used as an example of the multi-channel choice available to many American viewers. In 1984 the town had a population of only 4178 people and 44 stations via cable and satellite.

In 1977 a US Congressional Report said there were almost 116 million television sets and these were switched on an average of seven hours a day. In 1946 there had only been 7,000!

The mass of stations makes a review of any aspect of US television more complex and confusing because standards and practices vary so widely. There are however three main networks, NBC, CBS and ABC, which dominate and have coast-to-coast transmission. Their graphic design

*From Lewis Carroll's 'Alice in Wonderland'

* Television was built-on to commercial radio without debate'. 'Television – the first fifty years' by Jeff Greenfield. Published by Harry N. Abrams Inc New York 1977.

standards are high and similar to those of other countries with well-organised internal graphic design departments recruited from art colleges and more recently the new computer programmers. NBC (the National Broadcasting Company) was the first network and this was established by RCA in 1926.

United Broadcasting, another radio company, went into television as CBS, (Columbia Broadcasting) in 1929. They wisely employed William Golden as an art director and his work on the corporate identity and on much of their programme output around 1951 is still in evidence today. The third major television competitor, ABC, (American Broadcasting Company) launched into video in 1943.

Due to false starts in technical standards and the use of the anti-trust laws of the late 1920s which were invoked to prevent monopoly of the sound waves, and later control of television, and its commercialization, the establishment of the new industry was surprisingly slow.

These three networks have spent a great deal of time and money on corporate identity – very necessary with such proliferation of channels – and they were among the earliest to marry electronic stills stores and digital paint systems to make graphic video workshops for news programmes.

The arrival of ITV
Commercial television in the United Kingdom – called ITV or Independent Television – did not start until nearly twenty years after the BBC began television transmission. This extension of the new medium to the public started at 7.15 pm on 22 September 1955 and it was heralded with warnings from many sides. The Archbishop of Canterbury said this intrusion into our homes should be prevented 'for the sake of our children'. Some people still feel the warning should have been heeded! To divert attention from the ITV opening night the BBC 'killed-off' Grace Archer in their serial 'The Archers' in a fire in Ambridge. Was this diversion planned or chance?

In spite of this melodrama the first television commercial featuring a tube of Gibbs SR toothpaste frozen in a block of ice, was seen by a very small audience of only a few thousand people in black-and-white on tiny 8 and 14 inch screens. (The method is to measure screens diagonally –

maybe to give the purchaser the best impression). At this time graphic design was in its infancy.

There is no doubt that crude transmission determined the style of graphic artwork and film animation. The examples shown on pages 49 and 50 exemplify the heavy lines, the bold type, the extreme contrast of tone, and the lack of detail which was the sign of the times. Much of the work of the 1960s reflected the 'Op Art' of the period, and the technique of 'click' animation-cutting from one image to another in rapid succession – also dates the graphic animation of this time. (See Armchair Thriller on page 56).

Technical standards and systems
Compared with film projection, or the printed page, the television picture is still very coarse. When the moving image is 'frozen' as a still frame, the true quality of the line structure can be seen. Even using the present day British standard of 625 horizontal lines to the depth of the screen does not produce a resolution to match 35mm film projection. Twice the number of lines will be needed to equal this.

The screen image is made-up of hundreds of thousands of dots and in colour television groups of three dots carry the red, green and blue separations of the RGB signal. These points, or pixels, (smallest picture elements) are illuminated by a flying dot controlled by a signal from the transmitter. The whole of the 625 lines are scanned 25 times every second in the British, or PAL* system. This means the picture changes 300 times in 12 seconds using two fields per picture.

The standards vary throughout the world. In America the standard is NSTC** and this operates at 30 frames a second to match the 60 cycles per second of their electric alternating current output. This results in a steadier and less flickering picture, but has the disadvantage of fewer lines – only 525.

In 1950 a group of scientists, representing the countries throughout the world who were then developing television systems, met in New York to discuss the possibility of a universal transmission standard. They failed to agree and there are now the three systems, PAL, SECAM*** and NTSC, worldwide.

*PAL – Phase Alteration Line System.
**NSTC – National Television Standards Committee.
***SECAM – Sequential colour with memory – France and the USSR.

How does that effect graphic designers 35 years later? Electronic equipment for graphic design and computer-aided animation would have been simpler to use and cheaper had they been successful and video tape standards could have been universal. International co-operation at an early stage could have improved the picture quality for everyone.

Picture resolution
The British system improved from 405 to 625 lines in 1964 in the year when an additional Channel was granted and BBC2 was launched. This improvement allowed much more detail to be used in all forms of graphic artwork and presentation and higher standards of film animation began to be employed. The tolerance in the appearance of lettering was improved and by the late 1960s the very clumsy lettering of the first years slowly disappeared.

Into colour
The introduction of colour, the most dramatic step forward, came in 1969 on both ITV and BBC. Other countries had already begun colour transmission but the spread of this advance did not depend on the networks but on the number of viewers who could purchase colour receivers. The cost was high and even by 1970 when there was a choice of three channels with colour, in the UK there were still only about 200,000 colour sets compared to a million monochromes.

Better receivers
The next stimulus to gain larger audiences, which has been a continuous process, was the improvement by manufacturers in the quality of the television receiver. Larger screens and more reliable electronics have spread the ownership of television sets in Britain to over 18 million, of which only 3 million are black-and-white.

Computers and film animation
Ten years ago computer-control was introduced to the rostrum cameras. These were then the main source of making any form of animation in television graphic design. Effects like streak-timing and slit-scan became widely used because the computer, harnessed to the rostrum, allowed hours of precise and repetitive movements to be made, well beyond the ability and control of a hand-operated camera unit.

Painting by numbers

Television graphic design and animation have benefitted in a spectacular way from the application of computer science over the past decade. Few viewers are unaware of its impact.

'Painting by Numbers' was the title of an excellent edition of the BBC 'Horizon' series and it told the story of the emergence of computer technology – highly developed in the United States for the space programme and then applied to the world of entertainment. Yesterday's three-dimensional graphics for flight simulators became on-screen graphic animation for television idents and programme titles by the mid-1970s.

The emergence of the graphic designer

Throughout the years the broadening of the service has had the effect of encouraging bigger and bigger audiences. As these grew, the income grew from licence fees, and in the case of commercial television from advertising at a spectacular rate. A television franchise was 'a licence to print money' said Lord Thompson– perhaps unwisely.

John Halas wrote in 1967: 'As the output of graphic design for television reached considerable proportions, more and more stations established their internal graphic departments. The departments have grown in size and some of them, such as those of CBS and NBC in America, and the BBC in Great Britain, now have a staff of up to 40 persons.'*

The first graphic designers

Richard Levin was one of the first professional designers of his generation to work in television. His experience, prior to becoming Head of Design at the BBC in 1953, was mainly as an exhibition designer and like many in the ebullient days of the 1951 Festival of Britain was important in enhancing the role of designers and design in many new areas. Animators, typographers, illustrators and designers for print, as well as those trained in architecture, all gained a great deal from the Government sponsorship provided by the Festival. Recognition of design was then followed-up by industrial backing – all part of the post-war recovery.

Levin was able to persuade the BBC hierarchy that there was a rising generation of designers who were well-trained and that there was an increasingly necessary job for them to do in television. They would plan the presentation of stills, the production of lettering and, however simple at that period, make animated sequences. In 1954 he arranged for the BBC to employ their first full-time art school-trained graphic designer. His name was John Sewell. John had been at Hornsey School of Art where as a student he was seen to have a strong interest in film and film-making which he carried with him to the Royal College of Art's School of Graphic Design. This was then under the Professorship of Richard Guyatt, a contemporary of Levin, and also a contributing graphic designer to the Festival of Britain. British graphic design for television had found its first footing.

Was this far-sighted activity at the BBC prompted, or inspired, by the news that by autumn of 1955 the rival network of commercial television would appear to woo the audiences away from the twenty year old BBC monopoly? Whatever the cause, the move was well-timed, shrewd and in the long term very influential.

Training for television design

From 1954 the number of television designers began to expand at the BBC and in ITV. In London ABC and Rediffusion, in Manchester Granada, in Birmingham ATV, and all the smaller stations employed graphic designers who adapted to the new medium very quickly and later moved on to other companies.

Most of them came from art schools where, since the Second World War, the emphasis had been on applied design rather than the 'fine art' dominance which had pervaded them pre-war. Change towards industrial and applied design occurred in all major countries.

The strong influence of the Bauhaus and industrial design was beginning to spread through the 1960s and permeate art education. The Central School of Art Graphic Design Department, under Colin Forbes, and the London College of Printing, among other schools in the UK were very influential in removing the bias against 'commercial' design, as well as providing the students who could be recruited into advertising and television work.

Another contribution from John Halas from 1967: 'Film and television graphics have suffered from a feeling of inferiority in that they were outside the mainstream of production output. The flow in the case of television has in fact become a flood with a tendency to sweep away the finer points of production, where the role of graphic art lies. But gradually *even in television* (my italics) good design is emerging.'*

Animation in the early days

The impulse to produce moving images was very strong but few directors realised then, or now, the amount of work necessary to produce the type of animation which they saw in the cinema and early television commercials. Teams of people are required to produce even short amounts of hand-drawn animation and the more elaborate each cel the more work there is for the tracers and painters. The graphic designer working in television alone, or with perhaps one assistant, has always had to find 'shorthand' methods and use as much improvisation and ingenuity as possible.

The first years set the pattern for graphic designers to work on more than one programme at a time. Set design, and other major disciplines in television production generally demand the full-time commitment to one programme or series at a time. Research, design, modelmaking and supervising the building of a range of sets for a major drama series entails weeks or months of work, even with assistants. The vast amount of work and the number of animators and cel painters required for even a few seconds of animated film were outside the range and cost of making titles for most television programmes seen only once by a few thousand people. This limitation resulted in a very marked style. The rapid succession of a few still frames – and by using 'scratchback' where a drawing, or artwork, is slowly masked or painted-out, then shot frame-by-frame on film and run in reverse to build-up a complete image. (See 'City '68' page 50).

Graphic personalities

These are some of the changes which have taken place since the era of the smallest screens – when television receivers were 80% furniture and barely 20% screen – to the present day when our televisions are 95% screen and operated by remote control. The next section concentrates on individual designers who have made important contributions in this period.

*Film and TV Graphics/John Halas/Graphis Press 1967

*Film and TV Graphics/John Halas/Graphis Press 1967

Graphic design for television is youthful enough to have seen major changes in the traditional craft side of the business; everyone can feel they are 'pioneers'. However, some people and events stand out as significant and influential.

Three figures have been selected. **Saul Bass** because he was a catalyst for television design due to his inspired work for the film industry in the mid-1950s. **Bernard Lodge** is admired by his peers in Britain, and by designers abroad, as one of the most consistently creative graphic designers for television in the past twenty five years. He has never let his acquisitive and penetrating interest in *how* things can be done take over before he has solved the design problems. **Martin Lambie-Nairn** led the break-away from the internal graphic designers within the television companies in the UK and became one of the first to set up an independent graphic design unit working for ITV and BBC. This occured when the trade unions and the companies had matured to feel more comfortable and the importance of graphic design had become better established in the mid-1970s.

conference entitled 'The Changing Image' at The Barbican to address an audience mainly made up of young graphic designers, most of whom had not been born when he worked with Preminger and Hitchcock. One piece he showed on that occasion from his current work was a few seconds of computer-graphic animation and it still had the Bass 'magic'.

1 Saul Bass — making graphics move

The television networks in the USA were busy organising themselves well before the 1939-45 war but were no threat to the cinema industry until the mid-fifties when home television audiences began to increase dramatically.

One of the methods used to combat the decline in the cinema attendance was better publicity. Designers were employed to create not only the posters and printed material but the film titles.

Among the most influential and successful of these was Saul Bass. His designs were powerful enough to translate from the animations of the screen to posters, record sleeves of the sound tracks, and other outlets. Previously most film posters, with the exception of some specialist cinemas like the Academy in London, and the work of Jan Lenica in Czechoslovakia, had the subtlety of fairground flyposting.

Links with the film tradition

In this century film grew from a toy to one of the largest and most widespread industries. The silent film in the first twenty five years was an international medium, powerful because it worked in moving images largely transcending spoken language. Applying words, captions, and subtitles to a silent film was an early intrusion of the graphic element. The technical achievement of sound broke this polyglot and from then on every country had to develop separately. Masters of the silent screen like Chaplin and Buster Keaton could be seen and enjoyed with only the minimum of sub-titles by people in any country. when Al Jolson said 'You ain't heard nothing yet', in 1928, sound took away the intense concentration of

storytelling by pictures and the style of the directors and art directors of Hollywood who relied purely on images was diluted.

Television versus cinema

Using graphic designers for film presentation in the fifties sharpened-up the sense of the eye and design influence was very wide. Saul Bass made animated titles for a number of Otto Preminger's films. The first was 'Man with the Golden Arm'; then 'Around the World in Eighty Days' (which had a very long and amusing cartoon animation for the end credits), 'Anatomy of a Murder' and many others. These brought new creative energy to the mundane use of hand-lettering filmed over yards of draped satin — the cliché film title for many years. A carefully designed cinema title will help set the mood and even begin to tell the story, but it acts too late to sell seats.

The television opening title has a more direct job to do. The interest and the attention of the viewer needs to be gained and held. The repetition of a familiar and pleasing theme, using both sound and vision was found to be very effective. It is ironical that the cinema's effort to hold its audience by using graphic design was such an important stimulus on television graphic designers.

Many of the early television titles in England and America owe their style to Saul Bass. The use of photography in the 'Armchair Theatre' titles by Jim Gask produced for ABC in London and 'Famous Gossips' by Alan Jeapes for the BBC, which echoed the proportions of the newly available Cinerama system, both paid homage to Bass.

A little over thirty years after Saul Bass had designed his inspiring film work he came to London in April 1986 to a

2 Bernard Lodge — 'Lock, no hands!'

There was always a strong wish on the part of many designers and engineers to produce 'graphics' using video techniques. The idea was to eliminate film and painted and drawn artwork. To create new ways of controlling the television monitor to make images.

'Images generated purely through electronics have been with us just about as long as television. The world's first high-definition television station, at Alexandra Palace in London used almost from the very beginning as its test signal a pattern known as 'artificial bars' — a black cross on a white background — which was not formed by pointing a camera at a black cross on a white backgrounds but by an electronic circuit.'[*]

The familiar colour bars, are produced electronically, not by using art work.

Bernard Lodge was one of the first graphic designers to succeed in this objective. He was able to do it in a memorable and effective way in the title he designed for the first 'Dr Who' series made at the BBC in November 1963. (See page 54).

He exploited the effect which occurs when a video camera is directed at a monitor and the swirling cloud-like images produced were later the basis for the title of the programme. Essential to any effective title, and equally memorable on this occasion, was the soundtrack. The 'Dr Who' music was good and was developed electronically by the BBC Stereophonic Workshop.

Bernard had joined the BBC in 1959, after leaving the Royal College of Art, and worked there until he set-up his own company. This became Lodge/Cheesman in 1977 when Colin Cheesman, (who had been Head of Graphic Design at the BBC), joined him. They started at the time when film and computer technology were

[*]Rod Allen (Publisher of 'Broadcast' magazine) 'Images generated through electronics'/EBU Conference 1981

changing dramatically and when, as they wrote themselves, they wished to 'use technology not as an end in itself but as a key to open new creative doors'.

The other area to which Bernard Lodge contributed so much effort and applied experiment was in the development of film animation techniques using the rostrum camera without the use of conventional hand-drawn animation. He did much of this work with Filmflex Limited, about a decade ago, when computer-control was harnessed to the traditional rostrum camera.

Two main techniques stand out. *'Streak-timing'* where motion control on each frame produces an effect of drawn and blurred light; and *'slit-scan'* where the artwork, or transparency, is viewed through a very narrow slit (as small as 1.5mm). The slot is then panned, with long exposures, on to each frame while the camera is moving. This creates controlled distortion of the original and the movements are only limited by the patience and ingenuity of the designer and rostrum camera operator. (See page 108).

A great deal of the work created by Lodge/Cheesman was for commercials and for the cinema. They contributed graphic effects sequences to 'Alien' and 'Bladerunner'. Both would admit that their 'apprenticeship' was in television graphic design and they wrote, clearly distinguishing the difference in purpose between television titles and commercials.

'Title sequences often resemble advertising films, both having in common a relatively short time-span with the subsequent concentration of image and information. But a commercial can stand alone, whereas a title sequence is by definition part of a greater unit. It must 'warm-up' the audience and hold attention long enough to establish involvement in the actual programme. This must be achieved while remaining faithful to the programme's contents and spirit.' Following these tenets Bernard Lodge has produced a succession of apt and memorable television titles.

3 Martin Lambie-Nairn — an early independent
The bright colours of the Channel 4 company identity used on the cover of this book brought the design group Robinson Lambie-Nairn a Gold Award from D&AD. This, like most other 'overnight' success stories, was the result of many years of determination and hard work.

Martin and his company have enjoyed an enviable reputation in the design and trade press. One article on his progress was headed 'The life and TV times of a graphic wizard'.

He joined the BBC after studying at Canterbury School of Art (the same School, by chance, as Bernard Lodge) where he, like so many of his generation, came to grips with film animation as the basis of design for television. At 20 years old he joined Rediffusion.

When Rediffusion lost its franchise he transferred to LWT and served another 'apprenticeship' specialising in current affairs. He was there until the mid 1970s and established a reputation as one of the most proficient graphic designers but he also found himself dissatisfied by the second-rate position he thought graphic designers were receiving within the television companies. He chose to leave and teamed up with fellow LWT designer Colin Robinson in a general graphic design partnership. From the beginning print work and television graphic design were handled side-by-side. They prospered, moved to larger premises and from there wrote a canvassing letter to an executive at the newly created Channel 4.

The letter worked and in competition with some of the larger, and longer established, graphic design groups in London they won the day. The now very much admired flying sectioned '4' was produced in the United States because no computer house in London was capable of carrying out the animation. The whole exercise was thought to be extravagant but the proof that this was well spent and contributed a great deal to the acceptance of the newly founded Channel 4 is inestimable. This success has now brought confidence to a group which divides its output between 40% television graphic design and 60% print and general design. The television work has included a number of animations for advertising campaigns and given great credibility to the claim that good graphic design is a vital part of all television presentation.

An early film animation combining 'Op' and 'Pop' art elements of the early sixties. Designed by Roy Laughton, a pioneering member of the BBC Graphic Design Department. From his 1967 book 'TV Graphics'

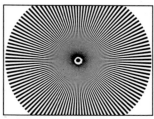
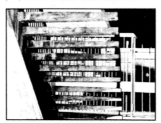

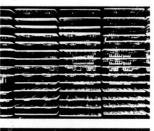

The need to incorporate lettering, symbols, still photographs, illustrations and moving images on the television screen has created a unique branch of graphic art.

Providing captions and titles was a basic requirement in the new television medium and from the very beginning all the methods known to the older branches of the graphic arts - drawing, hand-lettering, printing, photography, and systems of the early cinema – were used to originate words, and graphic images on the small screen.

After many years of discovering and combining ways of presenting visual material for video the process can clearly be seen to be a vital and integral part of television transmission. Graphic design can extend the range and express ideas and information in forms which cannot be achieved by any other method.

Vivid descriptions

'Graphic' firstly implies drawing, painting and writing, but a secondary dictionary definition *vividly descriptive* helps to show the purpose of applying time, skill and energy to the whole enterprise.

Illuminating the author's, or the producer's intention is the only possible purpose and justification for any form of graphic design. In following this aim the television graphic designer is part of the long tradition of calligraphers, book illustrators, engravers, lithographers, typographers and those who design for print.

A graphic designer can only aid communicators to translate ideas from one person to another, using images and sound, in an effective manner. To achieve this they need many, if not all of the skills of the earlier design applications; an ability to draw, as well as the imagination to create and combine all kinds of images; a knowledge and sympathy for the logic and history of lettering and typography.

Sound and music

The one distinctive element for the television graphic designer, is the ability to plan and produce images with *sound* and *movement*. Combining words, music and sound effects and then timing them precisely to pictures, frame-by-frame, is the essence of the craft, and this combination has proved to be one of the most compulsive ways of gaining attention. If this seems a little dramatic the simple

experiment of turning-off the sound from even the most powerful sequence of graphic animation will reduce the effect, or hide the mood and purpose, to an unaccpetable level.

An Open University programme (UK) on communication used the BBC titles for 'Secret Army' (page 95) to show how much the pictures depended on the soundtrack to give the key mood of tension and anticipation. When run with a more lyrical musical score the slow zooms to the vanishing point at the centre of the dissolving landscapes lost their poignancy, and reduced the strong opening sequence for a drama to one which suggested a travel programme, or a feature on country life.

Making movement

Although animation is a powerful weapon, it is, and always has been, costly and slow to achieve even the simplest results. The animator John Halas cited this point in his book, *Film and Television Graphics (1967)*. Writing on the future of film and television graphics he said the two main reasons why, even at that stage, graphic art was not more apparent in television programmes, were 'firstly the lack of awareness of the basic potential of graphic design among producers and directors who are the primary users of its services; and secondly; from the graphic designer's point of view, the complex technical processes which films and television have to apply before a final result can be achieved.'

Much of this present book indicates how technical change has sought ways of improving animation, many of them beyond the wildest expectations of twenty, or even ten years ago. However Halas' premise remains. Animation is still costly and complex. Controlling images, even the most simple outlines, at 24 frames per second is always going to be difficult.

The value of design

The typographic critic and historian, Beatrice Warde, wrote very eloquently about design and she had the clear philosophy that, in spite of her great appreciation of the art of the typographer, the reader should never for a moment ponder on the technical, or aesthetic problems of the type designer, or the typesetter. She thought of typography as 'the invisible crystal goblet' which would

reveal what it contained but not be apparent itself. Following this concept the television viewer should not be aware of the graphic designer. Graphic design is one of the few elements of television production which is involved in every single programme which is made. This point is not made to over-stress the importance of what is now almost universally referred to as 'Graphics', (avoid the ungrammatical 'Graphics Designer' please, this grates as much as 'Scenics Artists'), but to show how wide the application and the boundaries of the work of the graphic designer has become in television stations throughout the world.

Who does what?

In 1981 Michael Blakstad, then a senior and experienced programme director at the BBC, wrote a paper for a European Broadcasting Union conference on graphic design held in the Netherlands at Het Vennenbos. His theme was the role of the graphic designer and he said that he found it extremely difficult to define their tasks and he believed that confusion about their work was widespread. He was speaking to graphic designers from all the major countries in Europe and the intention of the conference 'TV Graphics '81' was to confront them with the emerging technical developments and to explore their effects on making programmes.

To highlight his accusation of confusion he cited his wish, when as Director of a BBC series called 'The Risk Business', to have the word 'Risk' in letters four feet high to dominate the set. The moment this was three-dimensional, and more than one foot high, the responsibility for its appearance became that of the Set Designers and *not* the BBC Graphic Design Department!

Who, he asked, could later say that the graphic designer did, or did not give the overall visual identity to that programme? He also felt that in most people's minds the graphic designer's job was, at that time, seen too narrowly, and he challenged the designers at the conference to be 'more thrusting and ambitious'. He would, he said, have been grateful as a programme director for more guidance from graphic designers.

The delegates from 44 broadcasting organisations and 22 countries found themselves largely in agreement with the main thrust of Blakstad's points in spite of the wide variations in the size, and difference of control, in their individual companies.

Things have changed since then. Partly due to the spread of electronic graphic equipment and computer-aided animation, the status and influence of the graphic designer in television has increased. About ten years before the Het Vennenbos encounter Richard Levin, the ex-head of Design at the BBC, made this prediction: 'I think that graphics may well become the most significant individual design contribution to television . . . The daily output of material for news, current affairs, children's and educational programmes alone using as they do captions, maps, illustrations, photography and drawings for animation is enormous; add to this the artwork required for programme promotions and it will be seen that the sum total of all this activity adds up to a large percentage of transmission time on every network'.

This projection has been largely realised. The rapid improvement and the better exploitation of the film rostrum camera by the happy marriage of the well-tried rostrum to the computer and the current explosion of change, effected by computer-aided design in so many applications, have enhanced the power and graphic presentation of programmes in almost every country in the world.

The daily work
Most television stations which make programmes – not all of them do – (Channel 4 in the UK employs graphic designers with the sole task of designing its on-screen promotions) cover a very wide range of broadcast needs.

A typical way the various interests are managed has been set out in the diagram opposite, although they will be grouped and organised to suit the management formula of each television station.

The range of work
Once graphic designers are allocated to a programme they work with the programme director as their 'client'. From then on he, or she, will be responsible for *all* aspects of graphic work for that programme, or series. This covers the opening title sequence, the credits, all programme content which could be anything from a single line illustration to many minutes of complex animation.

The graphic props within plays, variety shows and quiz programmes are also designed and produced by the graphic designer. These range from visiting cards, newspapers, posters, menus to large signwritten placards for location filming, or video recording.

Working with the Director
The relationship between the director, or producer, and the graphic designer is vital to the process of creating any worthwhile result. This formal relationship is easily understood and in an environment where there are a large number of specialists controlled by the director in a highly organised and very disciplined way, their inter-action is usually straight forward.

The programme director has an army of people to muster. These cover script writers, actors, set designers, lighting directors, sound engineers, studio staff, location managers, cameramen, film editors and VTR editors, property buyers and many more. With this number of tasks to manage the time available to become deeply involved in every level of production must be limited. The amount of care and interest invested in graphic presentation differs widely from director to director. To some it is very much an optional extra. To others the graphic work is a vital and binding element which can help to encapsulate the whole spirit and direction of the production. But as in any other human activity the 'best laid plans' do not always survive and expectedly good teams produce mediocre results while forced and sometimes unhappy parings produce work which lasts in the memory for many years.

Very few dirctors or producers working in television have any training in the visual arts. The output of the world of the moving images is in the hands of those educated to think fundamentally in words and literary terms. Their background if not academic is most likely to be from the technical or production side of television.

International graphics
There is a remarkable similarity of objective and working environment in the graphic design departments of different companies, and different countries throughout the world. A graphic designer working high up in the Rockefeller Centre on NBC in New York could change jobs with his counter-part working at the BBC on 'Thames News' London, or even, except for the language difference, with a 'graphiste' at Antenne 2 in Paris.

Watching a news programme in production at NBC in New York where a graphic designer was using a Quantel 'Paintbox' to superimpose the face of a well-known actress on to a photograph of an Egyptian princess, (to whom the actress bore a striking resemblance) was exactly like being in the Graphic Video Workshop at TVam in London.

Manufacturers of video devices, from VTR machines, stills stores, and those specifically designed for television graphics have all had to seek world-wide markets in order to make profits and survive. They are not helped in this by the different technical standards adopted in each country.

The British company Quantel has been very successful in designing and selling equipment to stations all over the world and their early digital effects machines were exported to Japan and America during the late 1970s. American and Canadian computer and digital effects equipment has been very important in Europe.

The graphic population
The number of people working as graphic designers in television has always been small when compared to the larger unrestricted fields of graphic design in printing, advertising, publishing and exhibition design. The reason for this is very clear. In most countries the power of broadcasting was seen from the very earliest point to be a threat to political stability and social standards and the control of this powerful medium could not fall into the hands of the 'wrong people'. As with newspapers, from their emergence in the 17th century, controls, taxes and licence to publish were established.

In the UK the BBC monopoly in sound broadcasting was extended to cover television transmission in the 1930s and this lasted until 1955 when commercial television, after many years of campaigning by a few people dedicated to this form of expansion, was finally successful.

It is strange that television was adapted to advertising support in the UK *before* radio. Sound broadcasting had operated in the UK for nearly fifty years, (from 1926 until the early 1970s) before the first commercial franchises for the airwaves were finally permitted. Even now the constraints over the commercial expansion imposed through the IBA (Independent

Organisation of a typical television company

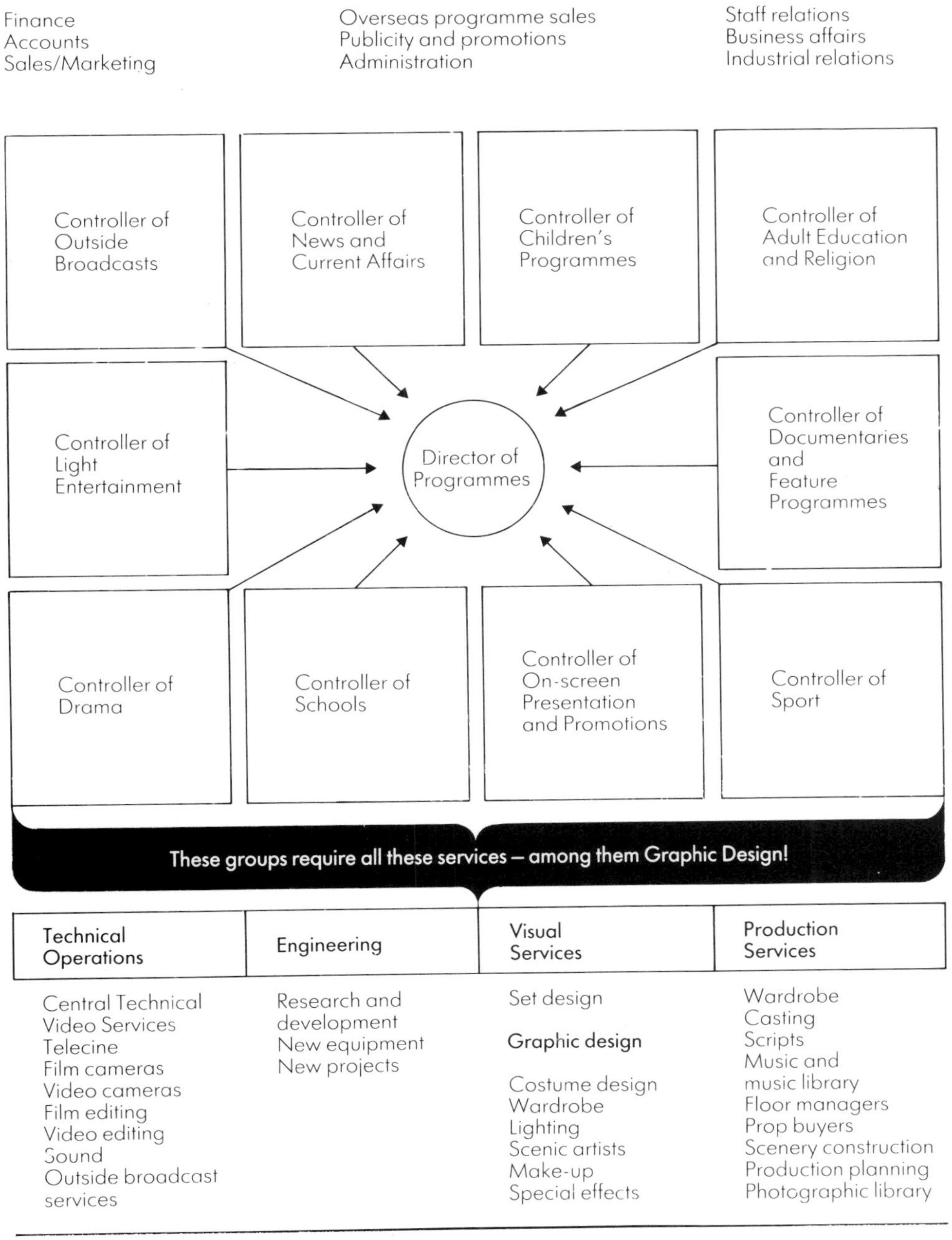

Finance
Accounts
Sales/Marketing

Overseas programme sales
Publicity and promotions
Administration

Staff relations
Business affairs
Industrial relations

Controller of Outside Broadcasts	Controller of News and Current Affairs	Controller of Children's Programmes	Controller of Adult Education and Religion
Controller of Light Entertainment	Director of Programmes		Controller of Documentaries and Feature Programmes
Controller of Drama	Controller of Schools	Controller of On-screen Presentation and Promotions	Controller of Sport

These groups require all these services — among them Graphic Design!

Technical Operations	Engineering	Visual Services	Production Services
Central Technical Video Services Telecine Film cameras Video cameras Film editing Video editing Sound Outside broadcast services	Research and development New equipment New projects	Set design Graphic design Costume design Wardrobe Lighting Scenic artists Make-up Special effects	Wardrobe Casting Scripts Music and music library Floor managers Prop buyers Scenery construction Production planning Photographic library

Broadcasting Authority) limit the number of companies given franchises, the hours they transmit, the areas they serve, and the amount of advertising allowed in any hour.

All these factors influence potential growth and change, as well as the output and type of programmes and in the end the number of technicians, including graphic designers, who can be employed.

In setting out these points no consideration is given to the merits of this, or any other way of using this quite extraordinary powerful medium. The estimated world audience for the 'Race Against Time', organised by Sport Aid and the United Nations Childrens' Fund, is more than 1.5 billion people via sixteen satellites. The 'Super Bowl' American Football programme went live to 130 million in January 1987.

At the beginning of 1987 the BBC and the ITV network employ around 600 graphic designers throughout the combined broadcasting systems. The number of graphic designers in areas outside television is vast in comparison.

How the graphic designer operates

A graphic design task in television can vary from preparing a completed single frame piece of information for a news programme in less than half-an-hour, to a two year full-time project on one series involving the control of a budget of many thousands of pounds.

A thorough understanding of the job is essential in both cases. In a situation comedy or drama series it means reading the scripts and gaining an insight of the director's interpretation. In news and current affairs it involves analysing often very sketchy pieces of information and doing a lot of research. The idea that graphic designers in television live in ivory towers does not bear much examination. Most are in the front line of production, or the subdued lighting of editing suites.

Animated sequences require a plan of action — or storyboard. These will vary from almost frame-by-frame realism to a barely decipherable scribble and a brief discussion. There is no perfect prescription — only the need to communicate the idea successfully. The idea is paramount. Once agreed the production methods are legion, although obviously constrained by the budget. Some attempts to classify them are contained in this book.

Personal style

It is not easy to establish an individual style in television graphics. The fixed format of the 3:4 ratio of the television screen, the tightly controlled production system in technical operations and editing, and the necessary large-scale teamwork involving many different skills – all tend towards conformity.

Television also imposes the time factor that everything every idea or message must be absorbed in seconds. There can be few pauses and no going back. A reader can move at his own pace and wait as long as he chooses before accepting the message and progressing.

Computer graphics have been blamed for the increased appearance of clichés and creating a bland international look. Reflecting metal surfaces and other effects – marvellous in themselves – have been applied too heavily and too often at times.

Some of the frustrations and limitations were set out by Scott Miller, Head of Design at WCBS New York, – 'The graphic designer's relationship with the medium is somewhat bizarre. The newcomer is usually someone who has been seduced by the energy, the electronic immediacy, the show business connection or the sublimnal thrill of knowing that perhaps millions will see his work in a single moment.'

Professional secrets

As in other professions there is great pride in discovering a new technique; cleverly marrying two old ones in an unexpected way; knowing just the right people to contact at the right time to get special types of work done; a favourite modelmaker or a rostrum camera operator who has perfected a technique.

This acquired knowledge is not usually zealously guarded. I have always found artists and designers very open and willing to discuss what they do with fellow professionals, other technicians, or enquiring students.

Budgets

The relative importance of a production process can be related to the amount of money required to provide it. What percentage of the total programme budget does towards actors' contracts? What percentage goes on scenery and set construction? A rough guide is that 10-15% of programme cost is spent via the production or set designer and only about 2% by the graphic designer.

Everybody needs an adequate budget to do the work for which they are responsible. The problem with graphic design's claim is that it is less obviously *essential*. If you are producing a studio series with a law court, various rooms in barrister's chambers and a scene in a café you can establish the likely cost of design, construction and painting of the sets. But what should the titles cost? And how do you know until there is an approved and well-liked storyboard? Higher expectations and increasing competition make most graphic designers feel they are not well enough funded to provide the finished product.

Visual reference

All the design services require reference material. In the larger television organisations there is usually a well-stocked and highly-organised access to a wide range of material - illustrative, photographic and documentary information, both historical and contemporary. Outside libraries are of course used but the speed at which some guidance is required – particularly in the early rough design stages – demands a good in-house library.

Graphic design for news programmes requires large photographic libraries. These designs are held as thousands of 35mm slides; the future will see them transferred to electronic stills stores.

Structure of a typical Graphic Design Department in a television company	
Internal Staff	**External facilities used by Graphic Designers**
Head of Graphic Design or Supervisor	Rostrum and Video camera units
Graphic Designers	Computer animation companies
Graphic Design Assistants	Video and post production companies
Rostrum camera unit	Modelmakers
Character Generator Operators	Animators
Photographic Technicians	Typesetters
Stills Photographers	Model animation studios
Finished Artwork Technicians and Lettering Artists	Illustrators
Signwriters	
Computer Operators and Programers	

The classic contribution to television graphics. An opening title for a drama series using the mix and fade potential of the film rostrum camera
'The Mind Beyond' BBC/UK
Graphic Designer/Bernard Lodge

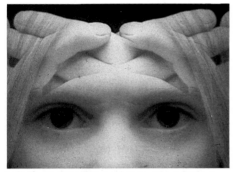

Over the last thirty years graphic design input has become more and more widespread and pervasive. Richard Levin's 1972 prediction that graphic design will soon 'add up to a large percentage of transmission time on every network' has been largely fulfilled. The surprise is that this took so long to happen in a medium that depends on images and pictorial information, and also that the value of graphic design was so late in being recognised and developed by many companies.

John Aston, (Graphic Design Manager at the BBC), remarked that the Corporation only began to recognise the extent of the contribution and the spread of graphic design when their costing system was revised in the early 1970s on the recommendations of the McKinsey Report. The report showed an unexpectedly high proportion of expenditure on graphic design.

Earlier parts of this book have stressed that graphic design is a *service* to the programme-makers and ultimately to the viewers. The objective is to translate the meaning and the ideas of the script into visual terms. The only purpose should be for the graphic designer working with the director or producer as an interpreter to enhance, emphasise and reveal the author's intention as clearly, and sometimes as dramatically, as possible.

Creative contribution
Accepting that graphic design is subservient to the will of the director – or ultimately to the Controller of Programmes – does not reduce the freedom to make a creative contribution and attempt to achieve the highest standards of production quality and inventiveness. Beyond a formal relationship the graphic designer must persuade and cajole, and in other ways influence, the director into accepting ideas and solutions.

If design becomes too strident, or more obvious than the content, then usually the point has been missed. While attending the annual conference on computer graphics – Parigraph '86 – I heard an experienced French television director, who had worked in British television, pay the following compliment: 'Britain is one of the few countries where television graphic design is still considered an art',

Roy Laughton, one of the graphic designers who joined the BBC in the 1950s, wrote in is book 'TV Graphics': 'The ultimate responsibility for the programme – the entire programme that is – sits squarely on the producer's, or the director's shoulders and it is he who has the final say about how his programme is to be presented in every department.' By 'every department' Laughton meant set design, make-up, costume, sound, cameras, script writing and the profusion of services which it is the director's task to co-ordinate. This edict remains unchanged, and unchangeable.

On presenting the same thoughts in the mid-1980s the slow evolution towards sexual equality makes the insistence on the male role sound too strident. There are now many women producers and directors within television; a point which is reflected in the increasing number of women graphic designers in the past few years.

The main areas
Although categories are never as tidy in practice as they are in theory the five main areas of graphic design contribution can be based on the following outline:

(a) The design and production of graphic material for **titles and end credits**.

(b) The design and production of graphic material for **programme content**. This covers stills, illustrations, captions, animated sequences and special graphic effects.

(c) **On-screen promotional material** for the television station, or network.

(d) The design and presentation of the **station, or network, identity**.

(e) The design and presentation of all graphic 'props' for studio and location set dressings.

Most graphic design staff will work on this complete range during their career. The extent to which they may specialise will depend on their personal interest and skills, as well as the way the company they work for manages the graphic design resources.

Section 4a Title sequence

Some television programmes are remembered for the design and mood of their opening title sequence years after they are made. The vast number of programmes produced does make the problem of preparing something fresh and original appear more and more difficult as the years go by. There is little worse than making a false start by planning a storyboard only to be told that the idea has to be abandoned because the chosen title is to be changed!

The power to decide on the amount of the budget and effort to be put into 'packaging'; that is giving a programme a graphic identity; rests with the producer and director through a Programme Controller.

Their attitudes vary enormously. Some will work very closely with the graphic designer from the earliest days of planning and rely on his skill and experience suggestions and advice. Others will try to dictate a formula or idea, often very literal, of their own.

The value of title sequences

'Graphic design and opening titles are always important but they are vital when you are introducing a series in which a lot of money has been invested. Titles can persuade people to stay switched-on – or not!'. This is the view of Pat Sandys a freelance director who has worked for both the BBC and ITV companies.

Another drama director, Tim Aspinall, said 'Titles are the overture and they should be a little play themselves. If the graphics generate enough impact viewers will stay with a show. We should never be parsimonous about the graphic design.'

But spending money is not always the answer. There is a dividing line between striving too hard for an effect and just being able to intrigue the viewer. One series was extremely restrained. The title for Granada's production of 'Brideshead Revisited'. The decision was to make titles using 'bookish' typography and rely on white lettering on a black background, plus the power of the music. Simple – but very appropriate – and in their way memorable.

The range of programmes

One of the positive elements of working in television graphic design is the extreme range of subject matter which is available. Even the very largest publishing house is unlikely to cover sport, news, the whole range of drama, from Greek drama to contemporary street theatre, childrens' programmes and light entertainment.

The examples of opening titles in this book can only be a tiny selection of the mass of graphic design which has issued from many television stations throughout the world in the past thirty years. From that point of view it is difficult to accept that they are representative. However they have been chosen largely to demonstrate the techniques described while some attempt to typify design and creative achievements which have inspired and influenced other graphic designers.

Some graphic titles

One landmark was Bernard Lodge's 'Dr Who' titles, made when he was at the BBC. They stand-out after over twenty years and generations of viewers recall the images of this short black-and-white sequence on hearing the name, or a few bars of the equally famous music. Later versions were made in colour to keep pace with the improving production values and changes in casting.

Alan Jeapes' always sensitive and yet direct solutions for the BBC Graphic Design Department have been remarkably consistent over many years. He has generally exploited film animation and photographic images rather than illustration and cel work in a very polished manner from his early 'Famous Gossips' (page 53) to the current title for the series 'East Enders'.

The 'World at War' titles were made for the Thames 12 part documentary series in the early days of colour transmission. The original storyboard by John Stamp and Ian Kestle is reproduced on page 57. Later the words of the title were altered and the memorial lettering and the sculpture of the soldier were removed from the final filmed version. A few clips from the final film are on page 51. The music, which carried a haunting theme to match the oppressed and tragic faces which are destroyed by flames was one of the early pieces to be written for television by Carl Davies. He wrote the music to fit the storyboard as well as the score for the entire series. Combining so successfully the music and graphic images for programmes like 'World at War' was important at the time

because it showed effectively – how much graphic decision could contribute to a major series.

One programme which gained a great deal of interest and publicity due to the opening titles was London Weekend Television's 'The South Bank Show' (page 78). The brief to the graphic designer was to create a feeling of accessibility to the arts, to avoid barriers and the often over serious approach to the arts and culture and to aim for the widest possible audience. The fast and unstuffy animations which have leaped to life regularly on-screen were presented to Melvin Bragg by The LWT graphic designer Pat Gavin back in 1977. Since then the design has been regularly updated by the same designer – a great tribute to the panache of the original concept. The music for this 40 second title was a recording by Andrew Lloyd Webber of the 'Theme and Variation' by Pagannini, and the music and the animations have now become inextricably inter-woven.

Model animation has been particularly well exploited on many occasions by Michael Graham-Smith of the BBC. His titles for 'Private Lives' and 'County Hall' (page 112) show how effective movement, the essence of television graphic work, can be achieved by constructing three dimensional models and filming them. To produce the result which Graham-Smith obtained in 'County Hall' could have been done using computer-aided graphics but the cost would have been very high for such a series. Many viewers, because of the highly reflective surface of the moving balls and the perspective of the chequered tunnels, may have assumed that they were computer work.

There was ingenuity in moving away from drawn animation and artwork by Pauline Carter of the BBC in her 'Match of the Day' titles of 1976 (pages 126/127). She borrowed the Chinese trick of gaining the help of a large number of people and arranging for each of them to hold in position a coloured card to form a gigantic picture. This title gained a lot of attention at the time, and helped the Sport's directors in television, and the graphic designers working with them, to be more inventive. It is a tribute to her and the programme makers of that time to take risks with time and money to attempt the unusual.

These few examples stress the importance of creative content and getting the message to the audience. The maze of techniques which are part of the graphic designer's armoury are discussed later.

The sound of music

The use of sound effects and music are vital. At best the graphic designer is able to offer storyboard with a complete idea before the decision on the music is made. A score can then be commissioned and composed to fit the accents and stresses in the storyboard. Sometimes the graphic designer suggests the music. This was done by Jim Gask when he was asked to design the opening titles for Thames 'The Diary of Adrian Mole – aged 13¾'. Ian Drury was commissioned to write the theme song to the storyboard. Only when there is a final version of the music, or the sound track, can the precise shoot, or the breakdown frame-by-frame be prepared. The same rule applies to all forms of animation and computer-aided sequences.

An example of an exception – they always exist – was the title made for Thames's drama series based on Alison Lurie's novel 'Imaginary Friends'. The storyboard was approved and the entire sequence was shot. It was all live-action with no optical work and then edited before the composer wrote a single bar of music. The graphic designer, Rob Page, had discussed with the Director, Peter Sasdy, his ideas for the soundtrack but the final result, the use of a female singer was a surprise interpretation to him.

The creative opportunities and the key part title sequences play in television presentation will always make them the most interesting and significant part of the graphic contribution.

Section 4b
Programme content

As in all other activities a vast amount of the work contributed is not obvious. Preparing graphic information and peripheral items required within programmes is just as vital and takes up as much, and sometimes more, time and effort than the design and production of opening sequences. Higher production standards and greater expectations from audiences have increased the quality and the quantity of graphic material within every type of programme.

Animated inserts

At one time a simple black-and-white map, or diagram, could be drawn and 'labelled' with transfer lettering, or Masseeley printed type, and set in front of a studio camera as a still caption. Such simple treatment is now often replaced by full colour animation lasting 10 seconds, or more, and timed to fit the commentary, or soundtrack.

For a long time graphic designers relied on the film rostrum camera to produce these animated sequences. An example of this is shown on page 61 where film animation using drawn and hand-painted cels, was applied to a documentary programme. Information about the Japanese economy was presented to the graphic designers and this was translated into a graph. To make this come to life, and to have a strong relation to the subject matter, the outline was ingeniously linked to a typical mountainous Japanese landscape.

The brief from the programme makers for a sequence like this might be given only a day or two, before transmission. The ideas and rough storyboard, the planning of the animation, the painting of the cels, the rostrum camera shoot, processing the film, and the editing would all have to be carried out in a matter of hours – often with late-night working and overtime to ensure the deadline was met.

The time needed to shoot and process the film also meant that current affairs and news programmes with their even shorter deadlines could rarely use animation of this type. A search for alternative and speedier ways of supplying programmes with graphic animation has been going on for years. The first move was to replace the film camera with video recording and this system is discussed in Section 6c.

Now small computer work stations are available and these offer the graphic designer working alone without a computer-programmer the possibility of preparing full-colour modelled animation within two or three hours. 'Alias 1' produced by Research Incorporated of Canada, 'Cubicomp Picture Maker' by Ampex and Symbolics are examples of systems now promising 'instant' computerised graphic design.

Many of these are already cheap enough to revolutionise the production of animated graphic programme information.

Illustration within programmes

Brian Trigidden, the Head of Graphic Design at the BBC, said this about drawing: 'Technology heralds a new generation of imagery but it is just another tool of the trade and whatever new techniques it may produce, computers still do not have a button to press called 'ideas'. Moreover if you don't have the ability to draw, you won't be able to do anything with a computer. If you put in a bad shape you will get a bad shape out. The ability to see the shape is born with the ability to draw.'

The level of illustration, skill and the variation of style of drawing varies immensely from one designer to another. Many can produce illustrative work to match the very best freelance illustrators and they could readily earn their living in that way. Others are not as fluent and hence tend to solve design problems in other ways, or commission outside illustrators. Improved transmission has allowed more and more detail to appear on-screen and enabled the style of drawing to be more varied and more subtle.

Few childrens' television programmes can afford full animation and when they do plan a full-scale production the project becomes so large that it is bound to be sent to a production company where there are many animators and dedicated rostrum camera operators.

Thames Television's associated company Cosgrove Hall is a good example. They have produced the 'Pied Piper of Hamlin' as drawn animation and an adaptaion of 'Wind in the Willows' using model stop-frame animation. Projects of this size could not be undertaken by a graphic design department within a television company.

Children's programmes, especially those

for the very young, use a great deal of illustration. Brief stories are usually presented without animation using only ten to twenty still drawings. For the pace and comprehension of the small children these are very effective and charming work is produced. They may be prepared by staff graphics designers or commissioned from freelance illustrators.

A further 'hidden' part of graphic design involvement with programme content is typified in this illustration prepared for the Thames Television documentary series 'Hollywood'. In the episode on stunts and stuntmen – an explanation was required to reveal how Harold Lloyd was swinging from a flagpole many feet above a busy street. A still from the film showing Lloyd grasping the pole was the starting point. Then, after much picture research, photographs of the period showing buildings and traffic were found and the montage shown here was created. A reconstruction of the set, as it never appeared in the film, showed the ingenuity of the early film makers in making the trick less dangerous than it originally seemed. The amount of re-touching and drawing in an illustration of this kind is easily overlooked when the graphic designer's contribution is so often thought to be confined to title sequences and end credits.

Production of such an illustration could take as much as ten days. There were 13 episodes in the Thames 'Hollywood' series and the graphic designer was engaged on the programme for just over two years.

Information graphics
There are many occasions when the graphic content becomes the major method of presenting the programme. This happens in documentary and feature programmes where there is a great deal of historic or scientific information to portray and the material does not exist in any other form, or previously prepared documents and prints have to be adapted to television presentation.

'The Body Machine' – a twenty six part documentary produced by Goldcrest Films with Channel 4 and Antenne 2 used a large amount of computer-animation to demonstrate many of the functions of the human body. The graphic content ws as much as 5 minutes in each 25 minutes episode.

Special graphic effects
From Drama to Light Entertainment – scripts often require graphic designers to contribute visual effects. A character may touch a light socket and the glow around him, to highlight this action, can be added in post-production by graphic design using film mattes or video animation. Fantasy and science fiction often require the combination of extreme scales. Massive insects crawling among tiny human figures, or model robots in a real landscape may call upon graphic design.

All these can exploit the possibilities of the old film 'tricks' – aerial image, moving mattes and masters as well as techniques like Chromakey and Ultimatte which are part of the newer video technology. Planning and producing such effects is yet another graphic design contribution to programme-making.

The discrete nature of much of programme content work answers the question which often occurs when viewers see the graphic designer's credit at the end of a programme and wonder what he or she actually did!

The complex photographic reconstruction (above) is indicative of the vast amount of work that is not identified as 'graphics'. The rostrum camera pulled-out slowly from the first hold to reveal the ingenuity of Harold Lloyd's stunt in the 'Hollywood' series. (Below) An off-screen shot of one of dozens of post-production graphic effects of flashes and lightning that helped to make a Tommy Steele children's programme come to life.

'Hollywood' and 'Quincey's Quest'
Thames Television
Graphic Designer/Barry O'Riordan

Section 4c
Station and network identity

The 'look' or design style of any organisation is important in mass communications especially in areas of intense competition. For a system whose existence is based on sending pictures through the ether to millions of viewers its own appearance, or identity, would seem to be of paramount concern.

Striking a balance between the banal and the strident to create a symbol that has to be repeated many times each day, often over many years, has proved a difficult task. One problem is to reconcile the effectiveness of an on-screen animation with the requirements of printed and other publicity material. Another has been to be too rigid. Where consistency has worked well in print, display and packaging design the moving image has always had the capacity to have variation and constant but controlled changes.

TV Globo, a very large network in Brazil owned and developed by a family newspaper company, recognised and exploited the possibilities of not fixing its 'trademark' too firmly. In the early 1970's they designed a whole series of animations. Their graphic designer was a young Austrian, Hans Jurgen Donner, who had made his way to Rio de Janeiro and decided Brazilian television needed good design and he could provide it.

Few stations in the world have used graphic design to greater effect. TV Globo spent a lot of money, but they have spent it well. They were among the first television networks to use computer-aided animation on a large scale when, among others in the USA, The New York Institute of Technology, the Utah School of Engineering, and some pioneering facility houses were able to offer computer-aided animation that had not previously been used in television graphics.

Design management
Designing a corporate identity usually attracts the attention of a large committee to approve the proposals. Committees, large or small, have rarely been conducive to positive and imaginative solutions. Isambard Brunel, the great engineer of the nineteenth century, made a presentation to the Bristol Bridge Committee and on the acceptance of his noble design for the famous Clifton Suspension bridge wrote:

'Of all the wonderful feats I have performed since I have been in this part of the world I think yesterday I performed the most wonderful. I produced unanimity amongst fifteen men who were all quarrelling about the most ticklish subject – taste.'

The early television idents
Many of the very early television idents were clearly not designed by professional designers – there were none available! They generally evolved from the style of the cinema production companies, RKO, MGM and Paramount, and very few were dynamic enough to survive changing tastes and technology.

An exception was the bold and lastingly effective symbol designed by William Golden for CBS in 1951, and still in use over thirty years later.

A large proportion of television idents were designed before the stations went on-air. The first stations had to recruit graphic designers who hasd little or no experience of television graphics. Everybody was an apprentice.

Most of the early symbols originated from on-screen animations. Looking at them en masse there is little to distinguish them from trademarks of the period for engineering companies, banks or any other type of organisation.

The early British commercial station ABC had a franchise for the weekday programmes in London and Manchester and achieved a solution which lasted until they ceased transmission in 1968. A simple animation of three coloured triangles froze into another triangle and this worked equally well on-screen and in printed matter. These early television idents often reflect their state or national foundations. Shields and heraldry abounded. Each decade finds it impossible to live outside the graphic style, or 'handwriting' of its own period. Like most antiques, many symbols can be dated to within a year or two of their making.

Channel 4 breakthrough
How rare it is, and how pleasing, when chance, skill and inspiration combine in some strange proportion to produce a memorable and successful result. The creation in the UK of Channel 4 had these happy coincidences; a single number as a station name; a peak in the new computer technology as midwife; and one of the earliest independent graphic design

groups to be set-up outside ihe then established company structure. When first produced in 1982 Robinson Lambie-Nairn had to go to Bo Gehring Aviation in Los Angeles for the computer animation. No London-based facility company had the hardware, or software, to achieve what was envisaged. The symbol was designed to represent the diversity of the new channel's programme sources – all of which originate outside the company itself – and ever since its first appearance it has attracted wide acclaim.

Central change
The cost and difficulties of the large scale job of changing an ident were faced by Central Television in 1982 shen the Independent Broadcasting Authority decided to re-organise television transmission in the Midlands and lose the ATV image.

Central commissioned Minale Tattersfield, who in 1968 had designed the long-surviving Thames river skyline scheme, and selected them from six competing design groups. The new ident was a white sphere reflecting a rainbow-coloured crescent. From the start the aim was an infinitely changeable and progressive design rather than something too fixed and static.

The scheme was presented as a model which opened, presenting the rainbow, and the possibilities for 'surprising things coming out of the sphere as the years went by' was quickly seen. Central wisely restricted their panel of representatives to only four people. Geoff Pearson, Head of Graphic Design, his assistant Stuart Kettle plus two senior members of the Marketing department.

In their design for Thames fifteen years before Minale Tattersfield had been able to exploit the newly-arrived dimension of colour transmission. In this later corporate plan for Central they sought to suggest the emerging promise of satellite broadcasting. The scheme has stood up well and its application to printed publicity and stationery still looks good although change is now contemplated.

Symbols for satellites
New methods of transmission are creating new markets and emerging companies and networks involved in direct broadcasting by satellite (DBS) and cable broadcasting have turned to graphic designers to produce corporate identities for on-screen

presentation. There sometimes seems to be a perversness about the design process. As soon as a new method or technique is perfected after years of research and development, the designers who introduced them want to move back to older ways! But it is this constant wish to revitalise and keep changing that often marks out the best design.

Robinson Lambie-Nairn designed the ident for the satellite Super Channel first beamed in Europe in early 1987. The solution featured the word 'Super' with a multi-coloured bar beneath it. To present this they returned to traditional film animation at Snapper Films and Martin Lambie-Nairn commented: 'Most television idents now use computer graphics. They are very much influenced by American television and they do not show a lot of originality. Super Channel is European and therefore the American feel was wrong.'

Section 4d
On-screen promotion and graphic design

Television, like radio and newspapers, is in a very good position for self advertisement. The audience can be advised and enticed to view future programmes to be transmitted later the same day, or days, or weeks in advance – all within the same medium.

The perennial promotion caption.
Channel 4/UK

The number of graphic designers dedicated to this area will vary from network to network, and company to company, and some groups handling this side of television graphics can consist of six or more designers at any one time. It is certainly a major part of American networks and a great deal of money is devoted to this method of selling programmes, series and whole seasons of events.

When a station has invested millions of pounds in producing a series, or has purchased the rights to a sporting event – the Olympics for example – the need to build audiences through on-screen promotion method is very important. In the constant ratings battle to keep viewers, graphic presentation is a vital weapon.

Promotion animations
The overall design of promotion material usually re-inforces the housestyle, or the main symbol, of the station. CBS pays great attention to this aspect and the promotion sequence on page 73 by John Le Prevost, a design consultant to CBS, bears this out.

The main use of very brief, sometimes only 5 or 6 second, 'buffers' is to lead into and the out-of compilations which announce future programme details. As these will range from hard news, to a documentary on torture or a light musical comedy or chat show, the style has to be kept very general. Many chose a theme which reflects the idea of a miscellany. Rob Page's multi-coloured blocks for Thames Television, a span of time by Harry Donnington/BBC1, or Marc Ortman's space spectrum for Channel 4 (pages 70-71) are examples of this. Only the strongly seasonal associations required for Christmas, New Year and Bank Holidays and similar events dare to be more particular. If too specific their inevitable repetition over many days can be self-destructive.

Still captions for promotions
Still captions are the bread-and-butter work of the graphic designer. They have been a necessary part of on-screen information since television began. They are also produced to be used in the event of a technical break in transmission as an emergency holding caption, and are normally on screen for only a few seconds, yet their message must be instantly conveyed.

The volume produced by any one station in any single day's operation is high. Some of the early versions shown on pages 68/69 would have been transmitted from an easel in a studio. Later most were made into 35mm slides for projection through a telecine slide scanner. Now, as those of Channel 4, they can be prepared on a digital paint system and immediately transferred to a stills store for instant re-call and transmission in their hundreds.

Section 4e
Graphic set dressings, or 'action props'

The Production Buyers in television companies are extremely able at solving the most bizarre requests and finding the mass of day-to-day articles to dress studio sets and locations. However some requests are so specific that the graphic designers have to produce them: where a book has an invented title; a poster is required for an imagined event; menus are wanted for a restaurant which closed in the 1930s; or a new political party needs a symbol and a whole range of publicity; all these print and graphic items have to be organised and made to order.

The 'faking' and making cover an extraordinary range. Passports, tombstone inscriptions, letterheads, Roman seals, 'T' shirts for aliens from outer space, packaging to avoid association with brand names and street signs have all to be designed and produced. Most graphic design groups have sign-writers, graphic technicians or artwork specialists to carry out the finished work. The sheer volume makes this assistance essential.

The four episodes of the BBC series 'The Loves and Lives of a She-Devil' demanded an unusually large range and quantity of graphic props and some of these are shown in the designer's 'Graphic Description' of the production on page 27

For the drama series set called 'The Bill' many rooms in a typical Metropolitan Police station had to be re-created and the graphic designer at Thames Television spent nearly four weeks in preparing enough posters, notices on police duties and a mass of paper work to cover the notice boards in a realistic manner.

Not award winning material, but an essential core of television production.

Introduction

Section 4 classified the work of the graphic designer in television and this section sets out to give some insight into their daily work, and the way they solve the problems encountered. This has been done by asking seventeen designers from companies throughout the world, and one representative of a computer animation company, to speak for themselves.

The illustrations at the back of the book have been selected to give a broad spectrum of programme types as well as a mixture of techniques. They are divided between ITV, the BBC, and overseas contributions.

Graphic news

The new Director General of the BBC, Michael Checkland, said within a few weeks of his appointment at the beginning of 1987: 'News and current affairs are a core of activity which we must nourish.'

This suggested more and concentrated resources. Even the relocation of all BBC news and current affairs activities on a single site. Graphic design equipment, and personnel, are now a very large part of any move. The final contribution to 'Graphic Descriptions' concentrates on news graphics.

'Good Morning Britain'

Live-action title animation
TVam/UK

Ethan Ames, the Head of Graphic Design at TVam, describes the expansive and dramatic use of live-action in the titles he designed and produced for the start of the ITV breakfast station in 1983.

The design of the 'Good Morning Britain' titles posed various problems. The programme was still very much in the planning stage, and the company was new and untested.

Usually a designer is given ideas to work within by his director and then goes away to draw a storyboard. In this case I was given a fairly free rein.

What evolved was a series of discussions where myself, the Director of Programmes, the Programme Editor and two producers threw around a lot of ideas. There was no pressure on me to use any specific medium. There was a feeling that it should 'involve' the viewer, and appeal to everyone. I felt that to do this it should be a live-action 'real' and not 'graphic'. Computer graphics, the current vogue, I felt were totally inappropriate.

The music, by Jeff Wayne, was composed before any ideas for a title were decided. Discussions between the composer and graphic designer did take place during composition.

My original ideas included people and objects representing letters of the title; morning events like getting-up, breakfast, shaving, newspapers. Another idea was to reflect the aim of the programme to reach everyone in Britain by including people from various walks of life.

We came up with the concept of using large numbers of people to make up the letters of the words of the title I was encouraged to progress with this concept, though clearly some people thought it was absolute folly!

The first storyboard of the titles had the word 'GOOD' as a dawn shot of fifty freefalling parachutists linking hands to form the word. 'MORNING' showed sailors on an aircraft carrier. 'BRITAIN' involved masses of people in a field. The final image was of birds in the form of the words 'GOOD MORNING BRITAIN'. Many variations were considered; farmers ploughing words; bowler hatted gents flowing over Waterloo Bridge; a convoy of milk floats. It was then a case of solving obvious practical problems. Was the freefall shot possible? Could we get permission to film on an aircraft carrier? Where could one film a large number of people and control them? Should it be a helicopter shot or a crane? And how do you get pigeons to act? All these questions

Part of the early storyboard.

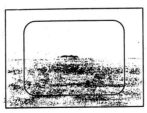

Sunrise . . . .

"GOOD" falls into screen . . . .

Zoom out . . . .

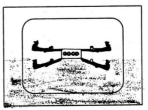

Keeping free fallers centred on the sun . . . . .

were a departure from the normal relatively solitary act of graphic design. They involved a lot of team work. I could never have produced titles of this scale without the assistance of enthusiastic producers and directors.

The completed title sequence is somewhat different from the original storyboard.

The original concept for the 'GOOD' sequence would have been very costly. We couldn't get 50 parachutists in this country and we would have had to go to the USA, where they're more experienced and have better weather! A trial jump was tried with a 9 foot by 3 foot banner with the letters 'GOOD' ironed on. The letters flew off at 10,000 feet. Then a strap with a thousand pound breaking strain broke.

All the sequences were shot in December and January, for first transmission on 1 February 1983 and the temperatures were unbearable. The 'GOOD' shot was finally accomplished with a banner four foot by two and a half foot and only two sky divers, and the result successfully started the title with a daring and dramatic shot, including the sun rising, a feature of the TV-am logo. The second sequence was to be pigeons spelling the word 'MORNING'.

All the pigeon experts thought we were mad. January is the breeding season and Trafalgar Square is somewhat depleted. Two-thirds of the birds are off doing other things! A trial feeding at 9.30pm indicated that the birds would congregate and feed off seed laid out in letter forms. Our first exercise in crowd control, was mainly keeping people (and birds) *away* most of the time.

When all the letters were 'seeded' the birds were allowed to feed. When the letters were completely covered in pigeons everyone made lots of noise and the birds flew off. This shot was reversed in the editing stage.

For the 'BRITAIN' sequence permission was obtained to use HMS Hermes and her crew. Having had the experience of the final sequence, which was shot *before* this one, and the advantage of the Royal Navy's training, this sequence was fairly straightforward. A graph of the word 'BRITAIN' was scaled down to accomodate the 600 sailors and the letters chalked out on the deck. Needless to say the sailors were perfect in their formations.

The idea of using a large crowd of people was transferred to the full title shot

'GOOD MORNING BRITAIN'. This was the most complex sequence to organise. A site was found at Durnham Downs in Bristol. Adverts were put in the local papers and leaflets were distributed.

The area was to be 212 feet by 117 feet, each letter being 31 feet high, and would require 5,150 people. Three hours before the arrival of the public, 80 stewards with 600 stakes, and 2,500 metres of yellow twine, pegged out the words.

Over 6,000 people showed up and the words were formed beautifully. Different shots had been worked out to cover as many options as possible (there could be no reshoot!).

A shot I had always wanted but that was thought to be impossible to control on such a scale was left till last. This involved the crowd breaking up and then, on cue, regrouping into the letterforms.

The helicopter crabbed around from east to south and then on a shout of 'Go' the crowd was to run to their original position. It worked.

In the final editing one clean shot, without any cuts, was used for each sequence. 'GOOD' lasts 6 seconds; 'MORNING' lasts 4 seconds; 'BRITAIN' lasts 5 seconds; and the finale lasts 7 seconds. An extravaganza of daring, movement, patriotism and people in 22 seconds!

Ethan Ames

This title appears on page 124.

The production of much graphic animation takes designers a long way from the drawing board.

Computer resources

'What is in the bounds of possibility *this week?*
Paul Docherty, Managing Director of Electric Image, — one of London's foremost computer animation houses — emphasizes the importance of a good relationship between television graphic designers and those who operate these rapidly changing resources.

In a company like ours, most of the operational personnel are design trained and have in fact worked as art directors or graphic designers before getting involved with computer graphics. The ideal way for a client (whether he be a director or a designer) to look at us is as a consultant resource — we work with complex visual effects all the time and we may be more familiar with a particular technique or new development than the client. What we can't do, particularly in terms of television graphics, is be aware of the exact idea and feel the designer and production staff have developed for each case. We can suggest alternatives and occasionally enhancements but it is up to the originator of the concept to decide if these suggestions are valid. Although these enhancements can often provide more impact and power to the idea there is a constant danger of getting into the 'Technology for technology's sake' trap – 'Let's use this effect simply because no one's seen it before'. Novelty can be a valid aspect of an idea but rarely survives as an idea in itself.

Given these considerations you can see that the best time to bring an idea to us is as early as possible, when it may be on the back of a napkin, or simply a rough description in the head of the designer. The most successful sequences we've been involved in have all benefitted from early discussion. Often a designer with whom we've developed a working relationship will drop in for a chat and kick around a few 'What ifs' – 'What if I wanted to do something like this?' or 'Could that be done this way?'. On a more mundane but just as necessary level there are the 'How muchs' – 'If I wanted to do something like this roughly how much would it cost?'. This can save a whole lot of work carefully storyboarding and presenting an idea that the budget just doesn't stretch to. It is also a means for the designer to catch up on just what is within the bounds of possibility this week – there are always 'for fun' projects going on, and I have yet to see an interesting technique or treatment developed 'for fun' that wasn't in the end used in a commercial project.

I think that the right way to approach the production of computer animation is to come to us with a strong idea and a reasonably open mind as to execution. If the idea has already been presold to the producer or director in exact detail then it may be too late to take advantage of new and interesting directions that may emerge – developments that could not have been predicted by the designer as they might not have been possible the day before. The idea that the designer wants to get across, however, must be firmly thought out and solidly in place or you run the risk of having the technology run away with it.

Computer graphics as a medium has a tremendous unexplored potential for creative expression but it must be directed for the same principles that govern any area of traditional design. It must not develop into a purely technical exercise – in the end, it is about making pictures.

J. Paul Docherty

Below: Drawings produced by the graphic designer, Mick Mannveille, to enable the letterforms forms to be digitized at Electric Image for the ITV promotion animation 'New for '87', (Storyboard page 58), and a highly-finished rough to illustrate Marc Ortman's precise rendering to the programmers at Cal Videographics. (See page 70)

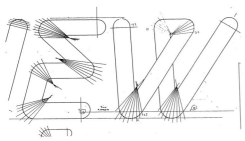

Three titles from TV Globo

'Magic Balloon' Live-action
'Fantastico' Live-action and computer animation
'Selva de Pedra' Model animation

Stills from these programme titles, each using a different production technique, appear in the illustration section of this book. The Head of Design at TV Globo, **Hans Donner**, based in Rio de Janeiro, describes how they were made.

'Magic Balloon' (Page 91)

We decided to use a balloon that has the power to give life, and colour, to everything it touches as the theme of this children's show. We built a circus setting where we were able to video record about 35 students from the Brazilian National Circus School.

They were aged from 6 to 14 years old. The children are trapeze artistes, acrobats, tightrope walkers, clowns and jugglers. Although they were all well disciplined and gave us excellent performances the total recording took eight days.

The circus tent was made with strips of multi-coloured cloth but in the master tape everything appears in black-and-white. The performers were shot against a blue background and Newsmatte was used to insert them onto the master.

This gave us three video recordings: the black-and white circus scenes, the colour scenes with the children and sequences with the balloon. They were edited together using four VTR machines and time code editing.

The climax of the title occurs when the balloon touched the highest part of the tent. This was the moment when we put colour into the whole circus using a wipe effect and the rainbow tent cloth came 'alive'.

'Fantastico' (Page 137)

In 1983 TV Globo's primetime Sunday night variety show celebrated its tenth anniversary and a new introduction was designed with flying pyramids and cones that were cut by rainbow rays to form the set for a modern dance ballet.

Initially we thought of using moveable sets with the station's main symbol, a huge sphere, but this was rejected because we thought this element was becoming over used. In the end we opted for pyramids — but these presented problems as well!

We produced 45 seconds of computer animation in San Francisco, and the high resolution work took 25 days to generate. During this phase Richard Chuang, the mathematician responsible for this part of the project, would report for work with a textbook under his arm. It was a pleasant surprise to find that the Institute of Pure Mathematics is not far from Globo's offices in San Francisco. Today the Institute is one of our sources of graphic computing personnel.

The other elements of the titles were produced at home in Rio de Janeiro. Musicians, dancers and the costume designers were all hard at work. Then we started building a giant blue pyramid in Rio's huge Maracana soccer stadium. The pyramid had five levels and measured 15 metres at its highest point. The idea was to put all 26 dancers on it at once. But something refused to fit between the 'eye' of the computer and the video camera; the real scenes and computer output just would not match. Meanwhile the dancers and 200 other studio and technical staff were waiting to start recording. The engineers checked their calculations, and tests showed that the idea would only work with the dancers placed on the ground. The pyramid came down.

We ended-up shooting each level of the pyramid separately. By chance Sony had just come out with the BWX, a machine that recovers colour and definition. We alternated these images with detail shots of the dancers. In the sequence where the set appears in front of the dancers we used masks made with a digital paint system. Some people thought it was the energy of the pyramid that was at odds with us. We even received letters telling us not to play with the pyramid. Some staff were involved in car accidents or quarreled with their spouses. Originally scheduled to take four days the video recording lasted 13 days. In all, the 90 second title took six months, or 5,000 hours to produce.

'Selva de Pedra' *(Page 116)*
The main character in this drama series, Selva de Pedra, has his roots in the countryside but he becomes a very powerful person in the 'big city'. The concept was to explore the reflections of the actor's face on the mirror surfaces of buildings. In the first sequence we see the light of the rising sun on cracked earth and broken buildings coming up through the ground.

The middle part of the sequence was made with 30 models and reflected photographs. Those scenes were shot in an area 100 metres square, with curved corners painted like the sky to avoid the buildings reflecting the studio. The camera movement made the clouds appear to move and this gave more realism to the shots. The lighting was done from underneath to avoid direct light on the buildings. In the final zoom-out the camera and the set were both moving at the same time.

To form the character's face there were 2,800 buildings in three different tones painted from a three tone photograph as the reference. We had to project a reference on a monitor and we shot many versions before we had the perfect position.

At the final editing session the mix of two takes was so good that many technicians believed that this was a computer generated image.

The planning, the research for materials, and the construction of the models took one week working 15 hours a day. Post production was another week and the final product was ready . . . 20 minutes before going on air.

Hans Donner

'The Beiderbecke Affair'

*A live-action filmed title for drama
Yorkshire Television/UK*

To suggest a character without showing him and evoke the atmosphere of the series without revealing the plot was **Diana Dunn**'s task for Yorkshire Television's Drama Department. She decided to use live-action.

The brief for this series, from the Producer, (slave driver!) was a familiar one. To embrace everything in the scripts — but to give nothing away! To inspire me I had a set of gently ironic scripts, and a very clever pastiche version of Bix Beiderbecke jazz as the title music.

Synthesising the sequence of Alan Plater's thriller, which was built around a complex central character, seemed impossible to do with purely graphic images, particularly the newly fashionable use of computer imagery. I felt none of these techniques would fairly represent the earthy, chauvinistic woodwork teacher 'hero'.

After further discussion with the Producer I plumped for the so-called 'obvious solution' — a live-action film sequence. As every graphic designer knows, this always seems the easy option, but of course it is very dependent on the talents and skills of the rest of the production team, and forces the graphic designer into a 'Director' role.

I decided I would like to create images of our hero from his immediate background — a colourful and eccentric bachelor flat created on location by the Production Designer. I felt the selection of objects and vignettes chosen could tell volumes in seconds, and hopefully set the atmosphere.

Having obtained agreement on the visuals to be used at storyboard stage, by both the author and producer, and breaking-down* the music with the film editor, we re-created the necessary parts of the main character's room in the stills studio at Yorkshire Television. This gave us more scope to manoeuvre than the original location, which was essential as I only had one day to shoot *all* the material needed.

As always, the trick of trying to get the pictures moving and not just looking like a sequence of still photographs took a fair bit of juggling. The temptation is to go too fast, or too slow, hitting the pace that will remain refreshing and exciting to the eye for the six weeks that the series runs, is the most difficult part of any title sequence.

The lettering was chosen to have a 'period feel', without being so dated that the audience would anticipate a period drama, and the Producer was keen to use a roller caption for the end titles — personally not my favourite form of graphics! designed and shot as a background, a large close-up of the spinning record, a simple but effective device, that not only visually related back to the opening titles, but also echoed the heros obsession with jazz, and in particular the jazz of Bix Beiderbecke!

Diana Dunn

This drama title is shown on page 97.

This drama title is shown on page 97.

*'Breaking-down' music entails listening very carefully to determine precisely where beats and stresses fit the film frame-by-frame.

'The Old Grey Whistle Test'

Cel animation for rostrum camera work
BBC/British Broadcasting Corporation

Those who originally saw this exuberant example of television title making, where the music and the images interlocked in an hypnotic way, still remember its power. The graphic designer of this stylish piece, **Roger Ferrin**, recollects the background to its creation.

'The Old Grey Whistle Test' was designed at the onset of the 1970s, and shot on 16mm Eastman colour. From concept to completion it took 2½-3 weeks with a budget of approximately £250. No opticals were required as the whole sequence was shot in camera.

The quality and meeting the deadline were achieved only by the undoubted skill and co-operation of Bert Walker (Zephre Films) who used minimal artwork to maximum effect.

Because of scheduling, time became the great editor, and our whole approach in solving this problem was very direct. Spontaneity became the key note, and perhaps that is why it survived for so long.

The imagery was strong, basic and strident, leaving doors open for the onlookers' interpretation. As we all know, it is more rewarding solving a clue than completing the crossword.

In retrospect it still seems incredible that something designed in haste, on such a small budget, should stand the test of time.

Roger Ferrin

Stills from this title appear on page 51.

Note: **How the programme got its name.**
BBC Producer Mike Appleton was asked to plan a programme to review the ever-expanding pop music scene of the early 70s. The venture needed a new name. Many were rejected as predictable but from a suggestion by the programme's researcher, Gloria Wood, came the enigmatic 'Old Grey Whistle Test'.

When composers in Tin Pan Alley wanted reassurance that a few bars of a new song would be 'catchy' they called anyone in the vicinity into their office. After playing the piece once or twice the victim – often a grey-haired Soho veteran – would be asked to whistle the tune. If he could do so the 'old grey whistle test' had been passed.

Roger Ferrin's design became one of the longest-running of all television titles and it is strange that he eschewed all visual references to anything old, grey, whistling or tests! What is more important to the graphic designer – logic or intuition?

'Channel 4 News'

Computer-aided animation
ITN/Independent Television News/UK

New techniques have been applied to all types of programmes but computers came to news titles very early. High budgets and constant repetition of the titles coincide. Graphic Designer **Lesley Friend** applied the astonishing 'metal-look' to her ITN titles.

The titles were part of an overall revamp of the 'look' of the programme and the initial concept started in October 1984, was to cover set, titles, content graphics maps, typography etc. The 'high-tech' metallic look idea was developed at an early stage, and it took approximately a month to design and storyboard.

In May three computer graphics companies were shortlisted by general assessment over a period of time. They were Electronic Arts, Digital Pictures and Cal Video. Cal Video was eventually chosen because they had recently written the software for 'chrome'. An estimate of 4 to 6 weeks work was given, at a cost of £16,000.

Below: Stills from the previous 'Channel 4 News' filmed titles.

Below: Images from the graphic title were the basis of the studio setting

An initial idea was to have a 'camera' watching the animation of the titles, but it was decided that the modelling might be too complex. Therefore the choreography of the lettering logo was progressed, with some discussion about the merits of the use of neon effect to alleviate the possible excess 'chrome'.

After further technical discussion, and decision making by the programme Editor, a compromise was reached with the Designer, Editor and Cal Video technicians.

A final storyboard was made, and after approximately four weeks with minor changes being made, a satisfactory conclusion was reached and 15 seconds of chrome flashed on the air, at £1,000 per second, on September 2, 1985.

Lesley Friend

The ITN titles are on pages 82/83.

'The South Bank Show'

Cel animation
London Weekend Television

An influential title, admired by viewers and graphic designers, which seems to encapsulate the contribution design makes to television. **Pat Gavin** reviews his association with the design of this long running series 'The South Bank Show'

I had the inspiration of a spark arcing between God and Adam's finger-tips back in 1960, but I didn't think of it then as a graphic idea, nor as the solution to a problem. This is how it happened.

I was walking passed Zwemmer's bookshop in the Charing Cross Road. In the shop window was a book, on the cover two hands pointing at each other, almost touching. As I turned away I thought I saw a vivid blue spark jump between the finger-tips. As a lad of fifteen I knew nothing about the effects of peripheral vision or I would have realised then that what I had seen, in the corner of my eye, was an illusion. The hands are from Michelangelo's Sistine Chapel, the spark was a negative image of the cracks in the ceiling they were quite dark, almost black. A simple trick of the eye reversed the polarity from black-to-white, and a trick of the mind colourized it with my favourite colour progression – blue through to violet and positioned the spark between the finger-tips.

The spark became a friend, I thought of it often, like a good memory. A snapshot – one of a clutter of events and images that insist on staying with me. That was 1960.

Then in 1975 I was asked to design a set of titles for a TV arts programme that would be called 'The South Bank Show'. My main problem was to try and find a symbol that could express recognisable images associated with the arts, but none could do the job. I filled many waste paper baskets.

Melvyn Bragg, my Editor and Producer, had suggested something about Trade Union banners. As a basic structure this would give a series of frames that could contain illustrations of the various arts and as such was workable and would be something to fall back on; a solid 'rainy day' idea. Amongst the other ideas I was working with was using the back of a canvas on its stretcher frame. Because we can't see the picture on the other side it could be anybody, and of anything – good thinking. But how can you make something interesting from what is not there?

I also tried to use the tools of different art trades as giant objects flying over a nameless metropolis – paint-brushes, typewriters, grand piano's.

I wasn't happy with any of these so I listened to the music again. Sounds have always been a good source of ideas for me, I would even say my best stuff has been closely linked to sound, something about making sound visible. When I listened to the music this time, one note stood out as being completely visual. It was the down beat that begins the tune. It sounded very electronic. It was played on the electric guitar with lots of fuzz. It sounded like an ignition and I saw in my mind's eye a complete re-run of the spark on the book all those years before and I knew then that I had found the symbol.

The spark is the spark of life, of knowledge, of communication and of the creative art. On another level it represented the transmission of the broadcast signal, itself an electric extension of the spark of life, and all in eleven frames! The solution had been sitting there all the time. Only now it had found its problem to solve.

But problem solving is not the whole of it. I know how I would feel about a perfect solution that looked boring. In the end the thing has to look good, and making it look good is using new tricks. Something has to contain my plundering. I would like to think that it is a strong idea – So to me this is the most important part of the job. Get that bit right and all else follows. I think that 'The South Bank Show' out of all the work I've done is a good example of getting it right. I hung out for that one. I wasn't sure I was going to get it, but a gut feeling told me that something was there. It's just a matter of waiting until what you're looking for surfaces. I waited until the last feasible minute – until I was endangering my deadline – but I always knew that I was going to get it. Good ideas, being bloody-minded and flirtatious keep a bloke waiting. The hands and spark turned up as cool as you like with not a minute to spare.

The 30 second film was made in about three weeks. I worked with Jerry Hibbert and Dennis Sutton, sharing the animation and rendering between us. It was shot by Graham Orrin Rostrums and arrived an hour before the first show was recorded.

That was in 1977 – I have now completed six versions of these titles (still with the same music!!) – but all the ideas that I have used evolved from the first storyboard sessions. The reason they didn't appear in the first titles was because we couldn't afford them, but as more money was made available so these ideas would be used. It got to the point where these titles became an expression of what money can buy – a kind of 'cheque book graphics' and they were in danger of being too complex.

The current set are a return to basics. With the previous two titles I was using computer assistance, slit-scan and motion control to arrive at a final image. This time I wanted to do something that was based on drawing. Apart from the use of some moving photo-collage material, (Pete Townshend and Francis Bacon), the technique was identical to the first titles. So I have come a full circle. And because I have done so, I can now think in terms of applying my skills to the computer-generated image and finding ways of combining computers with the traditional ways, but only if they express an idea.

Pat Gavin

This title appears on page 78.

'Loves and Lives of a She-Devil'

A multiple technique animation and graphic design programme content
BBC/British Broadcasting Corporation

Designing for a bizarre storyline set in the world of publishing, and providing an unusually large quantity and wide range of graphic props, is described here by the BBC Senior Graphic Designer, **Michael Graham-Smith.**

'The Lives and Loves of a She-Devil' is a macabre satirical fable of a plain suburban housewife whose husband falls in love with Mary Fisher, a beautiful and successful romantic fiction novelist. The trauma of her marriage break-up unleashes extraordinary demonic powers inside her, and she uses them to transform herself and her life, and to exact revenge on her husband and his mistress.

The four plays in this unusual series demanded a large number of specially designed graphic props, such as book jackets, posters, letterheads, logos, lapel badges and signs which were to feature in the action, in addition to the 'pack shot' in a fictitious television commercial and a house-style for an employment agency called 'Vista Rose'. Being satire, we were licensed to parody the style of romantic fiction, for example, in our own book jacket designs for Mary Fisher's novels, which were all virginal white with title logos in pastel colours unified by the characteristically flamboyant gold embossed signature.

By the time I came to discuss ideas for the opening titles with the Director, Philip Saville, and Producer, Sally Head, I was already immersed in the mood of the series. We had discussed the possibility of creating the effect of the She-Devil's eye 'breaking through' the action at key moments during the plays to observe the results of her machinations. It was an idea which was subsequently abandoned, but was to re-emerge in stylised form as the evil eye symbol I used as a leitmotiv in the opening titles.

I researched cabalistic symbolism and discovered the frequent use of two potent images – the triangle and the eye – both of which clearly offered interesting

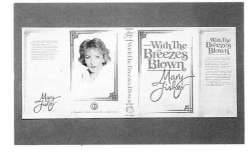

Part of the large quantity of books and publicity material required to establish one of the main characters as a romantic novelist.

possibilities for our theme. The triangle echoed the nature of the relationship of the main characters. The eye corresponded to the idea of the all-seeing eye of the She-Devil. I felt strongly that the eye should retain a symbolic, mystical quality, like a talisman – a real object, but with surreal power to transform itself. Around this germ of an idea I began to construct the sequence in terms of movement and additional imagery.

I had the idea of using live-action footage representing the elements of earth, air, fire and water as backgrounds, visual metaphors of the passions governing the lives and actions of our three main characters.

Such an abstract concept needed to be seen in conjunction with the characters themselves, and so portraits of the cast were shot on location – those of the two women were subsequently combined in the printing to create a metamorphosis from one face to the other. I now had a number of interesting visual elements, and it only remained to resolve the choreography of them to the title song 'Warm Love Gone Cold'.

I visualised the motion as one continuous tracking shot, drawing the viewer into the action like a vortex, tracking out from an extreme close-up of each character's eye to a wide shot revealing the portrait in a triangular frame, dissolving through the evil-eye symbol into the next close-up eye. Each portrait would be seen against an appropriate live-action background element, which in turn would add texture to the close-ups during the transitions. The final track-out would consist of continuous dissolves between the composite stills metamorphosing the she-devil's face into that of her husband's mistress. Philip Saville and Sally Head immediately recognised the potential of the idea I had storyboarded and presented and backed my judgement.

The She-Devil logo and evil-eye symbol were modelled, choreographed and rendered on a computer system which gave me frame accurate control over texture, lighting and motion. The rendered images and mattes were composited with the live-action backgrounds, still photographs and titles using the BBC's digital animation frame store and computer-controlled video rostrum camera.

Time, money and enlightened clients are essential pre-requisites for the production of good motion graphic design sequences – on 'The Lives and Loves of a She-Devil' I was fortunate enough to have all three, plus a talented production team who helped me turn my original concept into reality.

Michael Graham-Smith

Graphic Designer/Michael Graham-Smith
Assistants/Christine Buttner, Matt Anderson
Stills Photography/Conrad Hafenrichter, Peter Hobson
Live Action Filming/Oxford Scientific Films, London Scientific Films
Computer Imaging/Electric Image
Video Opticals/Peter Willis, Malcolm Dalton

The British Academy of Film and Television Arts (BAFTA) awarded this programme a nomination for outstanding achievement for graphic design, and the programme gained the Best Drama Series Award.

The opening title is shown on page 96.

'Crime Inc'

High-speed filming and photographic stills
Thames Television

Thames has made many documentary series. 'World at War', 'Immigrants', 'Hollywood' and 'Unknown Chaplin'. By chance they all had fairly monochromatic titles to suit their subject matter. **Lester Halhed** also chose classic black-and-white for this chilling six part documentary made in 1984.

My original concept for this title was to utilise the letters within 'Crime Inc'. These were to be cut out of metal with a bullet replacing the letter 'i' in 'Crime'. A series of images depicting criminals, murders, firearms, etc, were to be projected onto a curved metal surface giving them a surreal sinister distortion. A long slow track back, revealing the images, had in fact been projected on the metal head of the bullet. The track was to continue until the complete logo filled the title frame. The idea was accepted by the Executive Producer, John Edwards, and discussions commenced with the modelmaker and the cameraman. A bullet three feet in height was made and the remaining letters cut-out of brass. We began filming the distorted images with unexpected but beautiful results. Then came the first of several unforseen disasters – none of which could be rectified within the budget. An executive decision was made by John . . . I had to start afresh.

Idea number two incorporated a 'hooker' which opened every programme as part of the title. This action ended on a freeze-frame mixed to black-and-white and was followed by twenty frame cuts of fourteen more black-and-white photographs. This fast sequence of stills ended with a close-up of the face fo the Statue of Liberty which exploded signifying the complete break-up of American society.

My choice of music was just one bass note, beginning with the freeze-frame and ending with the explosion. This, I felt, would add a menacing feel to the title. However special music was composed.

The exploding effect was achieved with the image silk-screen printed onto break-away glass. (Sugar glass). Ten 24" by 18" sheets were ordered and carefully transported to the printers for screening.

I received a call from the printers a couple of days later assuring me that my fear that the glass would not break 'artistically' was totally unfounded. One of the printers had leant on the precious package breaking every single one quite beautifully!

More glass was ordered and screening began on its very uneven and knotted surface. Even a little pressure with the squeegee was too much for the delicate glass. Considering the difficulties the final results were remarkably good.

The live-action was shot at Stuart Hardy's with Doug Adamson on camera. An explosive's expert was hired to do justice to the now remaining eight sheets of glass.

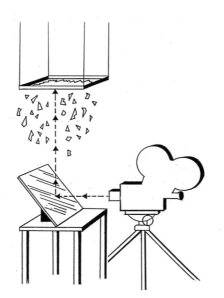

Using a mirror to film the exploding glass image at high-speed.

A Photosonic 35-4E high-speed camera (360 frames per second) was hired with a linked Polavision high-speed instant analyist camera. The glass was placed in a wooden frame suspended horizontally from the ceiling. The camera was directed at a silver surfaced mirror angled at 45° to the suspended image. The result was the face appeared to explode towards, and past, the camera.

There were several takes with varying numbers of explosive charges attached to the back of each pane of glass. None of which, when pictured on the instant replay camera, were to our liking. Two more sheets had broken due merely to the heat of the studio lights! By 6pm we still had not got the shattering effect we wanted. In desperation, five charges were attached to the last piece of glass. One in each corner and one in the middle.

The explosion was far too violent with the image completely broken-up by only the fourth frame. We had no choice but to make the best of what we had. The problem was resolved by double framing the whole sequence on the aerial image of the rostrum camera.

Linking-up the black-and-white stills with the live-action shoot was also done on the rostrum. Several versions of the mix from the explosion to the new title logo were shot – none of which worked as we wished. The answer suddenly became obvious. The exploding face, after a few seconds, should implode, forming the title. Another day was booked at the model-animation studio and this time the title logo was screened onto the delicate glass. Fortunately there were no mishaps on this second shoot.

The title explosion was then reversed on the aerial image – mixing in thirty-six frames to the end of the disintegrating face.

A touch of colour was needed to the otherwise monochrome title sequence. A 'tongue-in-cheek' splatter of blood on the logo was the result. This was shot under the rostrum on a black background and then supered on to the logo in a video editing suite. Title completed. QED.

Lester Halhed

Graphic Designer/Lester Halhed
Model animation/Stuart Hardy Limited Films
Screen printing/Rod Kitchen

The 'Crime Inc' titles are on page 80/81.

'Wales at Six'

Computer animation for a news title
HTV/Harlech Television/UK

David James, a Senior Graphic Designer, at HTV, does not believe in using computer realisation for its own sake. In this case, he says, the imagery could not have been produced by any other method.

The existing set and graphic style had been in use for some three years and had lost much of their effectiveness. In addition, the frequent turnover of Graphic Designers assigned to the programme (we work a 'tour of duty' system), had prevented a constant stylistic monitoring. This affects a visual style not only by different minds bringing different approaches to a predetermined style, but by newcomers to the programme allowing well-intentioned Directors and Vision Mixers to add their own contributions. The result is often an overall effect of design by committee.

The graphics and set for the new HTV Wales news style were designed in close liaison. We decided, at the outset, to dispense with the fussiness of elaborate decoration, logos nailed on to the set and an imposing graphic design style.

The set was to be a quietly lit space, using tonal gradation from left to right across the screen. Both set and graphics would use low saturation colours. Design, like acting, fails if it is self-conscious.

I considered, from the start of the design process, that the title sequence would serve a dual function. First, it is intended to set the style for Wales' national independent news and should suggest that what is to follow is of some importance. Secondly, it appears at six o'clock, a crucial switch-on point in the evening's viewing; viewers held at this point may well stay with the channel for the remainder of the transmission hours. It is, therefore, heralding HTV's evening broadcast.

The sequence was designed a year ago, to an open brief. The thinking and drawing-up process took one week, (although ideas were probably fermenting in the unconscious before this). The projected launch date was 1 March 1985. Digitization was continually deferred, owing to financial considerations with the set, which finally appeared in late 1986. This proved fortunate as I was able to take advantage of more advanced equipment, acquired by

HTV GRAPHICS DEPT. Telephone: (0222) 590356.
Programme Title WALES AT SIX
April 1985
Graphic Designer David James
Director/Producer Bob Symonds

'Wales at Six.' *Right: The line drawn storyboard of the animation presented to the programer at Crown Video Graphics. Below: The design link between the HTV station ident and their daily news programme animation.*

Crown Video Graphics, some months later, together with a new program devised by their senior programmer, Chris Fynes. It can be seen from the storyboard, that this replaced the rotating, revealing logo, in a cylinder form at the end of the sequence. Another casualty was the dragon on the front of the horizontal bars. The production office observed that it is an over-used symbol in Wales: as a recent immigrant, I was not to know this!

I was concerned that the sequence would date on the shelf and, indeed, other news and current affairs programmes are now using imagery with a similar hard feel. I do not believe that any conscious copying is involved: almost unconscious influences impinging upon designers from trends in taste, attitude and contemporary imagery are bound to throw out simultaneous, similar results. The same situation occurred when we came up with a new-look grid system for Thames Television News; and then everyone was using grids . . .

I chose to avoid well-used, pictorial, gritty, newsy images and decided upon an abstract, restrained, harmonious sequence of shapes choreographed in space: I

always design sequences as continuous actions if at all possible, such that each action sets the stage for that which is to follow. In this way the action, as well as the imagery, has a dynamic shape. You will notice that the feeling of the style acknowledges classicism, and this is enhanced by the use of only two colours. The logo livery is white type with the bars leading from pink to red from the centre but the effect proved unsympathetic when I tested it in this form in the sequence.

The complete sequence took three weeks to digitize, cost £7,500 and was programed in three sections which were edited together to match – albeit loosely – Amanda Alexander's music. As with music, animated sequences only exist as dynamic, temporal continua. The computerised sequence, in particular, is realised in a similar way to that of a piece of music: the images are designed in the format of a continuously-evolving sequence, but can only be realised – played out – by a programer, who adds his, or her, interpretation. In Chris Fynes I found a virtuoso performer.

David James

'Famous Gossips'

Rostrum filming in black-and-white using photographic stills.
BBC/British Broadcasting Corporation/UK

Alan Jeapes, a Senior Graphic Designer at the BBC, recounts his experience of his early days when graphic design was technically more restricted and the only way of making titles was some form of film animation.

Early in 1965 I was introduced to Patrick Garland. He was a young Producer with a reputation in theatre. I liked him. He was good company and receptive to graphic ideas. He was developing a series based on the biographical manuscripts of John Aubrey the 17th century antiquarian. Patrick Garland's television series was entitled 'Famous Gossips' and I recall that the sets by Julia Trevelyan Oman were superb. I was still new to television graphic design having only been involved in the business for about nine months and making the transition from print to designing for the moving image. Within weeks of joining the BBC I had made my first animated opening title comprising of spinning lettering. Bernard Lodge, who was already well established, asked me how I had achieved 'that great strobe effect'? I hadn't a clue. I then had some success with a title called 'Thorndyke', a whip panning, fast cutting piece. The sequence had no dissolves or opticals,* or what are now loosely termed, special effects. Frankly I did not know how to make them.

So it was that I started to work with Patrick Garland. Patrick had an idea of making a sound track using voices, whispering and gossiping in a mechanical rhythm. He employed the Radiophonic Workshop, the BBC's in-house group of composers, to manufacture this sound. I'd learnt that with a good sound track a designer can find inspiration and bask in the glory of the composer's work. On the other hand bad sound will ruin the finest design solution. With 'Famous Gossips' I was lucky. In homage to the poetic quality of the voices I decided to transpose the words to the screen in print thus doubling the impact. I photographed antique books as backgrounds and added a frieze of type to each scene using appropriate phrases from the script. In 1965 many designers were influenced by the work of Saul Bass and my frieze of type had something to do with having seen his titles for the film 'Nine Hours to Rama'.

'Famous Gossips' was a straight rostrum shoot with no opticals. The background photographs of books were overlayed with cels of the type frieze and dissolved one to another with a constant pan. The lettering in the centre of frame was double exposed on each scene during a second run. On the first shoot the lettering was not correctly exposed. At that time, when there was no electronic editing, a tape cut was literally a 2" tape joined with sellotape. Because this often resulted in a 'flash' the graphic material had to be played live to the master tape. In panic, with which one lived constantly in the early days and having no time to test the correct exposure for the type, I shot the background on pan (tone) stock, and the lettering separately on high contrast. Thus the first programme of 'Famous Gossips' had titles made by running two telecine machines locked synchronously together, the backgrounds on one and the central lettering on the other. The lettering was then superimposed using the studio mixer. The titles were reshot with the lettering properly added for the subsequent programmes. By sheer inexperience I had used a typeface with a combination of thick and thin strokes which needed critical exposure. Again, I was lucky. The panning background 'keyed' the edge of the lettering and gave extra legibility. When one remembers that black and white television had only 405 lines of definition and the clipping and matting systems were crude by today's standards some luck was needed to achieve any kind of subtlety.

Alan Jeapes

*Opticals are film processes made on an optical printing machine by a laboratory after film has been exposed on a rostrum camera or normal film camera. This requires duplicate negatives for mixes, fades, and superimposed lettering.

Stills from this title are on page 53.

'We Wish You Peace'

Live-action film sequence
CBS/Columbia Broadcasting System
Entertainment Division/California/USA

American television networks have commissioned outside design consultants more extensively than those in the UK and other countries. **John C. LePrevost** describes one of many animations his company has produced for CBS.

In October of 1985, I was asked by the Entertainment Division of the CBS Television Network to design ten seconds of animation to celebrate the holidays. The problem was to design something meaningful, contemporary and exciting.

After several weeks of careful thought and some rough sketches, we decided that the essence of the holidays was sharing, caring and, more importantly . . . peace. If we had one wish for the holidays . . . that wish would be for peace . . . therefore, the copy, 'We Wish You Peace'. The solution for the visual seemed simple . . . a white dove . . . a universal symbol.

Once the concept was decided upon, we began designing the storyboards and solving the possible production problems. The next step was receiving the necessary approvals at CBS.

The production began by hiring a top cinematographer. Next, we bought a pair of white doves from the local pet shop. We had no concept of how difficult they might be to control, how they would react to the handling and lights and how many times they beat their wings per second . . . which was the key to the timing of the spot. We were surprised to find that the doves were incredibly passive and easily handled. We were also surprised to discover the speed at which they beat their wings . . . an incredible 12 beats per second from lift off.

Once we understood some of these technicalities, the final production was a snap. We rented a high-speed Panavision (300 frames per second) camera . . . the services of an animal trainer and his doves and a sound stage with a blue screen. During filming, on several occasions, the doves escaped into the rafters, but the simple construction of a perch taped to a long pole with gaffers tape solved the problem. The birds simply hopped onto the perch and we began again. After about 7 hours and 24 takes we had finished. We had several takes that were acceptable and one that was perfect. We were all surprised at how close the final take was to the storyboard . . . it's as if the bird had seen the concept and then performed its part magnificently.

The next step was to combine the live-action (birds) with the animation (type) and then edit both with the music.

We all are pleased to have been able to work on such an interesting project and feel satisfied with the result. However, the true test of its success depends upon the audience's perception.

John C. LePrevost

CBS Executive Producer/Morton J. Pollack
Creative Director/Designer/John C. LePrevost
Associate Creative Director/Lewis Hall
Cinematographer/Gregg Heschong
Stage and Blue Screen Production/Apogge
Music/H&K Sound
*The storyboard for this title
and stills from the final filmed work are on
page 125.*

'Man and Music'

*Computer-aided animation
Lilyville Productions*

The mathematics of music and design. **Bernard Lodge** describes his use of computer graphics to create an harmonious visual effect. He set up one of the earliest independent graphic design companies in 1979 to specialise in film and television graphics – Lodge Cheesman – with Colin Cheesman, an ex-head of Graphic Design. Bernard had been a graphic designer at the BBC since the early '60s after he had left the Royal College of Art.

The commission to design a title sequence for 'Man and Music', a series dealing with the history of western music, seemed to provide an opportunity to escape momentarily from the field of computer graphics and return to a richer more traditional method of image-making. However, many of the visuals associated with music and its history had to be ruled out because the Producer, Tony Cash, justifiably, was anxious to avoid the obvious, predictable, musical images:

musical instruments, sheet music, portraits of composers etc. The chosen music, the opening bars of Bach's D minor concerto, also imposed constraints because it ran for only sixteen seconds and made a one-image sequence inevitable.

Examining the actual sheet music, I decided to go for a note-by-note synchronised animation of something that would have the abstract yet significant characteristics of musical notation. This developed into a spiral of 63 strokes, each stroke of different height equivalent to a musical note. At the storyboard stages it became apparent that only a computer could generate the semi-architectural image or cope with the visual counterpoint of quickly animating strokes on a slowly rotating spiral.

Vector, rather than solid, shaded raster imaging was necessary due to budget restraints, but the fine-line vector system used (IMI 2000) was in fact more suitable, providing an ethereal, musical quality, the fine vertical lines evoking something of the texture of 17th century copper-engraving.

Bernard Lodge

Graphic Designer/Bernard Lodge
Computer animation/Derek Lowe of Computer FX
Producer/Tony Cash for Lilyville Productions

Frames from this title appear on page 135.

'WCBS News'

Graphic design for news presentation

CBS/Columbia Broadcasting System/New York

Very large budgets are needed for all aspects of television news gathering and presentation. Because of this the news departments were the first areas to adopt video graphics, stills stores, character generators and digital paint systems. **Robert Scott Miller**, the Design Director for WCBS-TV in New York, described his station's design revisions for the new graphic era in an article in the Japanese magazine 'Idea'.

American television is going through a shape-up and shake-out. The audience is declining and profits are generally flat or

down because too many companies are trying to cut a piece of the shrinking pie. Those who seek to win viewers must cut through the entanglement of the video jungle. Differentiate. Establish unique personality, and real identity. Invisibility is the problem, measurable visibility the goal. Designers are now responsible for finding new and better ways to create visual separation between channels.

There is a different kind of graphic packaging needed today. Cutting through the clutter has become a priority for television executives. Graphic design has finally been acknowledged as one very important means to that end, and it is this change that has turned the world of the television artist and designer upside down.

At most television stations, the centre of graphic identity is the local news. This is the station's biggest moneymaker and its greatest opportunity to establish its image with the public. It's also the graphic designer's greatest opportunity to exercise creativity, as there is a continuous need for new work. But because television news is such a big business, and since a single slip of a ratings point can cost a major-market television station hundreds of thousands of dollars, the fear of innovation and change runs deep. Television news is so ephemeral, and the formula for success so transitory, change – particularly graphic change – is looked upon with suspicion.

A classic dilemma exists in this situation. To attract new viewers, a television station must risk offending and even losing existing viewers. The television news audience is considered quite fickle. Change of any kind is a gamble. Graphic change is particularly risky because it is so obvious and implies a corresponding change in editorial content even if none has occurred.

WCBS-TV is a New York City station owned and operated by CBS. It is held closely under the wing of its conservative and highly regarded parent. Yet, station management felt the need for innovation and a new graphic identity, and they decided to completely redesign the news packaging.

I don't know if anyone has tried to go quite this far with the packaging of television news as we did at WCBS. There were two major areas of concern, the news set and the graphics. Typically, these elements are designed separately, often by different people at different times. That shouldn't be surprising, as they are very

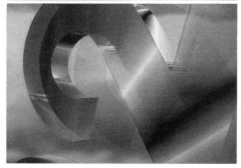

The opening title for 'WCBS News' Computer-animation – but avoiding spinning globes and views of cities.

different problems. The designing of a news set usually involves extremely complex logistical and technical problems, while graphics are a somewhat more flexible matter. Invariably, graphics become subordinate.

We did it backwards. Taking advantage of all that our new computer graphics tools had to offer, we determined that graphics should be considered first. We decided to ignore the traditional approach, taking the view that the news set was nothing more than an oversized, three-dimensional graphic in which people function.

Colour, of course, played a major role in creating the sense of continuity we wanted. But the most important aspect of the design

involved the architectural components. The design was based on real or imaginary three-dimensional forms including windows, plates, bars and gratings layered in such a way as to create illusions of dimensionality and depth. The viewer's television set is itself a component in that graphics are constructed to appear supported or reinforced by the physical dimensions of the cabinet and framework surrounding the screen. This consideration, however subtle, seems to play a role in the aesthetic qualities of the design.

The WCBS news set is based, quite literally, on our graphic design. Here graphic windows, plates and bars are displayed in real space as tables projecting from columns, walls and windows arranged to provide distinct layers of depth. Details include heavily cut geometric shapes, architectural reveals and a low sofit which prevents the design from soaring beyond human scale. Strong verticals are relaxed by long horizontal lines and rectangles which are compatible with the shape of the television screen.

The animated graphic used to introduce 'WCBS News' is also different. Traditionally, news openings have hyped the product with loud trumpeting music backed by the sounds of pounding teletype machines. The visual aspects typically have included spinning globes or helicopter-shot views of the city. A recent favourite is the lengthy animation which places the viewer somewhere in deep space, flies him past the moon; through the earth's atmosphere, over the city and down the street to the television station where he crashes through the doors of the studio and is deposited into the lap of the news anchorman who grins and reads the day's top story. This is fun to watch, but seems to risk the credibility of the news organization.

The 'WCBS News' opening, doesn't hype the news or attack viewers. Rather, it is an identity piece that seeks only to communicate a positive sense of WCBS. It is bold, yet it has softly coloured, transparent elements set against an 'idealized cloud' background. Response from viewers indicates it is effective.

Another difference in the WCBS design is typography. We decided to try to live without fat type. We wanted something with more style than the heavy typefaces commonly used in television news graphics. That presented problems. The medium of television is not kind to light

weight type and often chews it up mercilessly. To use light type, it is essential to provide a clean area for the type to rest. The WCBS design includes a narrow bar used to clear areas of the screen as needed. The bar doubles as a style element. But this is only a partial solution.

Most of the type used in a television news programme is produced electronically on a character generator. The quality of letterforms is somewhat poor and there is a limit to the number of founts that can be used during a live television production. In practical terms this means there may be as few as two or three type faces in only a few sizes. That's not nearly enough to manage the special requirements for using light type. (This consideration is one of the reasons you see so much heavy and uninteresting typography in this medium.)

WCBS started with the intention of doing something different. The station was willing to try things, take chances, it wanted to graphically separate 'WCBS News' from the competition. After several months on the air, indications are that the design is successful. Viewers write letters to comment on the style of the show – this is highly unusual.

But that doesn't mean we had an easy time getting this thing together. the list of contributors to the graphic redesign of WCBS was rather long. More than a dozen individuals, above and beyond the designers and graphic artists already at work, became involved in a project that included a complex news set with three major presentation areas, a number of computer animation pieces, and a very large battery of still graphic work.

One of the great challenges in the career of a television designer is to make management comfortable with something it needs to help establish the confidence of viewers. Credible and thorough graphic packaging. It's a hard place to get to.

As time passes, the imperative of visual separation and identity will become clear. New tools and improvements in the means of producing graphic identity, the availability of computer graphic techniques and anticipated upgrades in the medium clearly suggest that television, finally and forever more, is a designer's medium.

Robert Scott Miller

Graphic work for 'WCBS News' appears on page 84.

'Splash!'

A multiple technique animated title for a series of childrens' programmes Thames Television/UK

To make a sequence as polished and textured as this took many people and many weeks' work. **Barry O'Riordan** explains some of the problems and stages that occur after the someone says 'Yes' to the initial storyboard.

High-speed film, cel work and video effects mix in a mere 35 seconds.

A well-known television producer once compared the role of a graphic designer producing a title sequence with that of a father producing a baby – after the initial conception. Someone else does all the hard work. Lighthearted though this comment may have been, it does illustrate that even experienced colleages have failed to understand the true role of the graphic designer in television.

In making the 'Splash!' titles I was lucky in two respects. Firstly, in having a Producer who did fully appreciate and support the role of graphics and, secondly, being able to work with a small highly-skilled team helping me to produce the artwork.

The idea was 'sold' to the Producer, Kate Marlow, in a quick thumb-nail sketched storyboard; drawings that showed key points in the animation illustrating how the line was to develop.

My idea was basically simple – but complex to produce. With a title like 'Splash!' I did not ignore the obvious: The graphics were to grow around a series of slow-motion live-action splashes and the droplets of water, resulting from the corona fragmenting into images, were to cover music, fashion, environment and the arts.

After the thumbnail storyboard I drew-up a highly-finished production storyboard. This was essential, not only to help explain events to others who were to contribute, (the photographers, cameramen and the music composer,) but also to enable accurate budgeting of time and facilities., especially those that would have to be bought in.

The budget fought for and agreed left me facing the first problem – how to produce a series of live-action splashes that exploded over iconic images? I went to Oxford Scientific Films. They specialize in high-speed filming and there we experimented by dropping water on to photographic

transparencies. These were sandwiched between two pieces of glass to protect them from water stains.

Filming at 4,000 frames per second, spectacular though it may seem, produces its own unique set of problems. Was there any microscopic debris in shot? (Filming an area about 1" across meant that the tiniest fleck of dirt or dust would appear log-size on the film rushes.) Did anyone move during the few seconds it took for our 100 feet of film to speed through the camera? The slightest air movement would cause the water droplet to fall away from the ideal central point. Had we held focus? At the film speeds we exposed the depth of field was critical and how do you focus on such a small fast-moving object anyway? All this was experienced through dark sunglasses to protect our eyes from the high-intensity lights that we were forced to use.

Steve Downer, the cameraman at Oxford Scientific Films, overcame all. Applying logic, he 'froze' the speeding droplet with strobe-light, so that I could see to line-up my shot and then he found focus by suspending a metal nut, the same width as the drop of water, into position over the transparency. Three days filming and 1,500 feet of film later, I had the shots I wanted.

The next stage involved selecting and editing a cutting copy of the live-action splashes. An anxious Producer looked over my shoulder and breathed a sigh of relief. A Thames film editor, John Wright, added extra pace with some deft cutting, even at this early stage.

Now, thirty separate animated sequences had to be designed and animated, each particular to its own section within the titles. Twenty six were produced by myself and an assistant, Andy Newmark, two were animated by Reg

Lodge, and two by B.M. Animation Limited. The animations relied on a wide variety of techniques; conventional cel work; animating people and objects with still photographs; 3D clay model animation; optical and backlighting effects for a mock computer look. All of this was eventually shot with travelling mattes by Vic Cummings and Norman Hunt on computer-controlled Oxberry cameras in Thames rostrum unit.

The final stage of production entailed collating all the elements, live-action splashes, 35mm animation, digitized effects, and soundtrack on to one 1" video tape. Grant Williams at Moving Picture Company employed all his skill and patience in the final tape edit, running as many as seven tape machines, ADO and Mirage to produce the end result.

Now what was it that Producer said . . .?

Barry O'Riordan

The 'Splash!' title was exhibited at the first International Exhibition of the New York Art Directors' Club & D&AD the sequence appears on pages 90/91.

'The Innovators'

Computer animation for a feature programme
Airshow International Films

'Our biggest bit of hardware is a photocopier and it will stay that way – unless we get something like a Paintbox'. This attitude gives design companies like English Markell Pockett the freedom to employ any production technique they think appropriate. **Darrell Pockett** explains

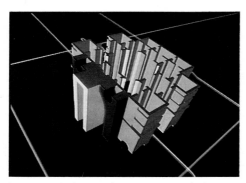

Innovation-doing what computers do best.

why computer animation was finally selected for this commissioned title. All three partners in this newly formed independent design company learned the graphic design business at the BBC.

'The Innovators' is a 36 second title sequence for a 26-part television series commissioned by Airshow International Films, initially for screening in the USA. The series showed the work of innovators in many different fields such as pop music, the performing and visual arts, architecture, fashion and industrial design.

The director's brief called for a visual interpretation of the word 'innovation', if there was such a thing, and for examples from the work of the contributors to the series to be shown. The sequence was also required to have an international flavour and an 'innovative appearance'.

When designing titles, or any other sequences, it is essential to identify the most relevant aspects of the brief, and in the case of 'The Innovators' the subject matter of the series was so rich in imagery it seemed natural to select this aspect as a basis for the sequence, linking these images with some sort of visual description of the word 'innovation'.

The whole concept was based around the original thought stressing the difference between the innovator and the mere mortal. It was the spark of an idea, the moment of inspiration which was fundamental in all creative endeavour. It was decided to use the device of a stylised oscilliscope trace to represent the thought process and link all the scenes with this graphic vehicle, activating the growth of appropriate objects on its journey. The culmination of the sequence was to be the trace line drawing the facial details of an otherwise featureless sculptured head.

Computer animation wasn't the sole choice of technique for the sequence. The first attempt at a storyboard combined live-action and computer animation but proved technically difficult and expensive to execute. The same idea was then tailored to suit a total computer animated look.

Like traditional film animation or live-action, the storyboard stage of a computer animated sequence is where the work is concentrated. It is here that the visual message, structure and style of a sequence has to be determined and accurately mapped out. Once the idea is committed to the storyboard it then becomes the most important means of communication between designer, client and animator. 'The Innovators' was particularly satisfying as there was little deviation from the storyboard to finished sequence.

The title sequence took two weeks to design and five weeks to execute. It was produced on videotape by a Vax 11750 main frame computer and cost a total of £22,000.

Darrell Pockett

Graphic Designer/Darrell Pockett
Computer animation/Cal Videographics

Frames from this programme title appear on page 134.

'What the Papers Say'

Model animation filmed with special lens Granada Television/UK

Model filming often has the ability to deceive the viewer about the scale of the objects presented. Here the reflective metal surfaces looked very large and suggested the strength and influence of the Press. Granada Graphic Designer **Peter Terry** reveals the three-dimensional lettering he used was barely the size of a playing card.

Not two feet but only two inches. That was the size of the lettering used in this model animated title sequence. The lettering was cut from ¼" thick aluminium, highly-polished and fixed to a black velvet background. The shooting of this sequence proved most entertaining at times, but the end filmed product was a close interpretation of the original storyboard showing various newspaper headlines reflected in what appeared to be random lettering until a track out reveals the titling – 'What the Papers Say'.

A very small model-but powerful when animated on-screen.

Having commissioned the cutting of lettering I selected a range of typical, but not too specific, newspaper headlines and these were to be reproduced 'back to front' and the 'wrong way round'. This would ensure when they were reflected that the wording would not appear as a 'Russian text' but be quite legible – as the programme producer had been promised – with fingers crossed!

The storyboard agreed, its contents constructed, it was time to arrange the shoot for the sequence. This was carried out on a computer-controlled rostrum camera – with the aid of an endoscopic lens. This somewhat resembles a metallic straw and is attached in the place of a normal lens, enabling the artwork, or in this case the model lettering, to be shot very tightly without losing focus.

The movement of the lens can be plotted via the computer, and this combined with a rotating, tilting rostrum bed can result in some very interesting effects. It is possible to view what the resulting image will be, before exposing it to film, by means of an eyepiece at the side of the lens – providing you are a double-jointed gymnast.

The lettering, mounted on black velvet fixed to a board, was attached to the rostrum bed. Above this in 'Heath Robinson' style the reversed headlines were attached to the camera itself with black paper and tape so as it panned across the lettering the headlines were reflected the right way round.

Various moves were set up for each separate headline and these were combined into the final timed sequence in an edit suite. To my delight and surprise the result came very close to the concept.

Peter Terry

The opening titles of this long running series are on page 76.

'The Italians'

Rostrum film animation using photographic stills
ABC/Australian Broadcasting Corporation

David Webster is a Senior Graphic Designer working for ABC in Sydney. He explains the pleasure he found in designing and filming the title for this documentary series.

The ABC TV Features Department requested a title sequence for a series of six 50 minute programmes documenting the history and experience of Italian immigrants to Australia.

The programmes included oral histories, archival film, stills and newsreel footage covering the political and social histories of the Italians in Australia. Much of the material, of course, was black-&-white and suggested a title which reflected and complemented this quality. (Several of the films began with black-&-white scenes).

In my search for imagery to use in the sequence a friend loaned me a catalogue from an exhibition of Futurist photography and montage which had been held recently in Italy. This approach with its juxtaposition of the reviled classicism and the new mechanical age seemed perfect for any monochromatic ideas. It also invited a dynamic animation treatment.

It remained, however, to sell this somewhat idiosynchratic approach to the producer.

This proved to be the least of my problems as he had a longtime fascination with the Futurist movement in Italy. So I was on my way and left very free to collect and collate material and formulate a sequence. A piece of music was settled upon at the time – an instrumental section from a Verdi opera which began gently and built

steadily to a crashing finale. It broke up roughly into sections which corresponded to the groups of images I had selected.

I decided to utilize a thin coloured line as a counterpoint and a thread at this time. I reanimated an old photographic sequence of a conductor, made Boccioni's sculpture walk again, and even managed to include the time honoured spinning headlines technique. The second last scene was an aerial-imaged piece of archival film of Italians boarding a ship to Australia, which brought me to the point at which the story really began.

I thoroughly enjoyed the whole exercise – a job in which I was able to employ a range of tricks and techniques which I had long been wanting to put to good use.

It was filmed on an Oxberry Rostrum on Eastman colour negative. There was an attempt to grade for the black-and-white on film and this was further enhanced on videotape in post-production.

David Webster

This title is reproduced on page 80.

Graphic News

Most television networks and programme contractors are involved in news presentation in some form. Headline news (Cable News Network – CNN) in Atlanta, Georgia are dedicated to news only. Whilst others like the BBC and the American networks CBS and NBC, have large separate groups of graphic designers dedicated solely to news output.

'Radio with pictures' was a jibe about news presentation for many years. What was the value of television when all we were able to see was a news-reader with a script in his hand wearing a bow tie – merely a 'talking head'?

When television began, the difficulty of getting pictures from one part of the world to the other was very different from the present situation.. To send crews and get film returned was dreadfully slow. Now live pictures by satellite can be transmitted from almost any part of the world.

A similar dramatic change has occurred in graphic design. Through a computer with a database map of the whole world any part can be selected, framed, and enlarged to the size required. The outlines can be

coloured and textured as required, the place names and other graphic information set in electronic typeface. All this in a pre-planned format to suit the programme can be ready in minutes. Using paint and hand artwork the task could take hours.

Until graphic information (still or animated) could be produced very quickly, the integration and importance of graphic design in news programming was bound to be held back. For a long time even in this picture medium graphic information was an optical extra.

The Director of Graphic Art at CBS in New York, Ned Steinberg remarked:

'Although speed is essential in all aspects of delivering the new the tools of an artist used were developed for a much slower industry – print. In fact, there was nothing electronic about early news graphics, except that ultimately our work was put under a television camera.'

The search for new methods began on both sides of the Atlantic about ten years ago.

ITN London

The ITV network of 19 separate companies in the United Kingdom pool their resources to pay for a central news gathering and production service. This is called 'Independent Television News' and each company has transmitted the newscasts from ITN since 1955. An additional task has been handled by ITN when they undertook the evening news programme for Channel 4 in 1982. They now employ 10 graphic designers and 10 computer operators to design and produce the constant flow of visual information seven days a week.

Their head of Electronic Graphics, Peter Atkinson, explained that since 1986 the graphic work has been 100% electronically generated. The build-up to that position had started in 1974 when a VT30 computer was purchased. This had limited capabilities for information graphics but it was used in the 1974 election – the first use of raster graphics in British television. The images it produced were like Teletext with very jagged edges but it had the enormous advantage that everything could be played back very quickly which is essential when large amounts of data are flowing almost continuously. Since ITN engineers, working with the graphic design group, have made their own system – the VT80. This was ready for the 1980 American Election and improved the appearance of graphics on-screen.

The graphic designers take their brief from the reporters and then plan and sketch out their ideas for animation which are then processed by one of the VC80 operators. A complex moving diagram, showing the operating faults of a nuclear reactor at Chernobyl was planned and ready for editing, with the commentary, within two hours. All this has to be contained 'in-house' because a close relationship between the editorial staff and the design team is vital.

Stills for both ITN and Channel 4 news are created and drawn by graphic designers on three Quantel Paintboxes at present and more are anticipated.

They have Harris stills-store and this allows simultaneous direct access to stills by any one of a number of designers, and by the main control room.

CBS New York and the electronic workshop
When the first digital paint systems and reliable stills stores were both on the scene the concept of an 'electronic graphic workshop' became a reality. 'Instant Graphics' arrived at CBS in New York over eight years ago when the Ampex AVA, one of the earliest paint systems, and an electronic stills store (ESS) were married. The stills store could record up to 800 images on each disc through a television camera mounted on a stand and retrived instantly by a computer terminal. The input could be from small colour transparencies to large colour prints. With eighty discs the CBS electronic library contained over 20,000 graphic stills.

Any of the same images could later be incorporated in the design of a caption for the news, via the AVA.

The electronic age in television news graphics had arrived.

News at the BBC
A build-up in research in computer presentation precedes each General Election and these periods have resulted in improved standards of work and co-operation between editorial and visual services. The BBC's long history and strong relationship between Engineering Research and Graphics have produced very valuable contributions.

They developed a digital paint system called 'Eric' which was later developed under licence as 'Flair' by Logica some while before the Quantel 'Paintbox', and many other systems, got underway.

Introduction
The sub-title of this book promises a review of the techniques used in television graphic design. These encompass a very wide range of image-making and control. Technical methods are rarely thought of, or used, separately as the following notes attempt to describe them. Television designers can incorporate many of them in a single item, whether it is an opening title or a graphic animation within a programme, and there is a strong current trend to mix the techniques. 'Splash!'/page 91 and 'Ghosts in the Machine'/page 79 are examples using multiple techniques.

From the earliest days, television called upon all the methods of originating graphic images known to printing, the cinema and other branches of the graphic arts. Television graphic designers have always been involved with every form of printing, from lithograpy to silkscreen, all methods of typesetting, and every conceivable way of making images. Drawing, painting, engraving lino-cuts, photography. The move to electronics and computers as a means of generating images for television has been surprisingly late. Only in the past five years has the graphic design studio seen much change.

For example, electronic character generators have been available for many years, as a means of putting lettering on-screen, yet many stations throughout the world still find use for hand-set typesetting on a hot-press Masseeley. Johann Gutenburg could still earn a living with his 500-year-old press in some of the best equipped television stations.

Over the years the main equipment and techniques are as follows, and a description of each is given under these headings:

6a	Film animation
6b	Model animation
6c	Video rostrum systems
6d	Colour separation
6e	Stills stores and digital effects
6f	Live-action
6g	Digital paint systems
6h	Computer-aided animation
6i	Lettering systems

Only a summary of each is possible in the scope of this book. The amount now written on computer animation alone is staggering.

Section 6a
Film animation

The film rostrum camera contributes to a number of areas of television graphic design production. The most obvious is the filming of animated opening title sequences but filming compilations of stills, — photographs, engravings, paintings or specially prepared artwork – for the content of every possible type of programme is still a very important function. Television stations, mostly those with large graphic design groups, notably the BBC, Granada, Thames Television and Swedish Television, have for many years had their own internal rostrum camera units while other companies have relied upon the skills of independent rostrum camera companies who also service the feature film industry and the makers of 'commercials'.

Video recording has not made this older system of recording images obsolete; although video has the advantage of immediate replay while the material is being recorded and mistakes can be discovered and corrected rather than having to wait until the film is processed by a laboratory.

Film has the advantage of being able to achieve fades and mixes of great complexity, and multiple exposures, with relatively simple equipment. The rostrum achieves three specific jobs. Firstly filming stills – typically press cuttings, photographs, or 35mm slides. Secondly, shooting cels and drawings single frame for animation. Thirdly, making compilations made by tracking and panning on artwork, maps or other illustrations and stills.

From cinema to television
This important tool of television graphic design was inherited from the film industry of the '30s. From 1928 onwards Walt

Disney had organised skills and taken hand-drawn cel animation to a stage of excellence unlikely to be surpassed. His full-length feature films 'Snow White', 'Bambi' and 'Fantasia' contain scenes which have so much movement and detailed animation that the number of people employed to prepare the thousands of cels can only be dreamt about now. This achievement in cartoon film refined the use of the rostrum camera and trained many people in its operation firstly to the advantage of the film industry then to television in its infancy.

Cel animation
'Cel animation is fundamental to all advanced cartoon film technique. The idea of using transparent cels may seem simple enough, but it was in fact a revolutionary one.'* Foreground drawings traced on to thin sheets of transparent celluloid and superimposed as independent images on painted backgrounds was patented in 1915 by an American film-maker, Earl Hurd.*

Using either 16, or 35mm film each frame of drawn or photographic, artwork can be shot at 24 separate frames per second. If enough different drawings are made the control of the apparent movement can be so smooth as to deceive the eye into accepting continuous action, either fast or slow, and approach the appearance of real movement. But total naturalism is not feasible and in drawn animation some form of stylisation is necessary to enable the illusion to work.

Rotascoping is an aid to the animator when the natural motion of human, or animal, is planned. Live-action film, or video, can be traced frame-by-frame as a guide to every movement and then transferred to cels. The rendering can be in any style chosen to suit the subject and the skill of the artist. Television graphic designers have often used this method very effectively and avoided the involvement of an animator.

Stills compilation
Rostrum camera animation can be used to bring still pictures to life. By panning, tracking in various directions, zooming in and out at various speeds to match spoken commentary, or musical score, a

*'The Technique of Film Animation' John Halas and Roger Manvell/Focus Press/1959

succession of images can be compiled and enhanced by film editing. Shooting stills in this way is not a substitute for cel animation, or live-action. Well designed and expertly shot it can enhance, explain or dramatise any idea or mood.

Whole title sequences are made in this way for television. Short films for childrens' programmes, documentaries – where only still archive photographs, maps, and engravings are available – and many news and educational programmes constantly benefit from large quantities of this form of filmed material.

An early form of animation in television graphics is hallmarked by the jerky 'click-animation'. Only a few changes in the drawings or photographs were required, and this labour-saving effect was widely used.

The rostrum system
The minimum requirement of the rostrum camera is to expose cinema film one frame at a time. To obtain the perfect registration of one image on another on frame-after-frame of as small as 16mm passing through the gate of the camera – pass after pass – demands extremely precise engineering on all parts of the rostrum and the camera. A photograph of an Oxberry camera, one of the best and most highly developed of its kind, appears on page 105. Backgrounds, artwork and cels are fixed to the rostrum bed on registering peg bars and the bed moves north, south, east and west, and revolves. The camera moves up and down the film column. All these can move to a single thousandth of an inch at a time and even when operated by hand the performance was very impressive. Fades from light to dark, and vice versa, as well as very slow mixes from one image to another can be carried out over as many as 120 frames lasting about 6 seconds. Making multiple exposures on each separate frame of film can reach a very high number of passes and beautiful effects of mixed images evolved.

Computer and camera
The rostrum camera, involving the co-ordination of many simultaneous operations, was an ideal target for computer-control. About ten years ago the successful marriage of computer and rostrum, with motors on all moving parts, was made in the the UK and the USA. The improvements in graphic design animation

were considerable. The potential to exploit the variety of moves with far greater speed and accuracy due to the computer's infallible memory, was opened up. Work, which even the most patient, careful and experienced rostrum camera operator could achieve in days, could now be carried out in hours – once the computer was programmed.

'Slit-scan' and 'streak-timing' were two techniques which became relatively simple. In these and similar techniques a single piece of artwork, or a colour transparency, which could be back-lit, is enough to create a sense of three-dimensional and highly complex movements, and distortions. The power of the camera does the work; the need for complex and multiple artwork is reduced. With cel work where the illusion of movement is built-up by many drawings the addition of a computer is of very little benefit.

Aerial image
This very useful adjunct to film animation allows film to be projected through a 45 degree mirror on to a ground glass screen on the bed of the rostrum. Cels of the drawings can then be positioned exactly where required on the projected sequence and then shot frame-by-frame. This is the technique used to combine cartoon figures with live-action, or two or more sources of live-action. Mattes or masters (Painted or cut-out cels) may be required to mask out foreground and background and follow the movement of the camera and objects.

Aerial image has been exploited by graphic designers in television on many occasions for similiar work, and for special effects. In one case the outside of a London theatre was shot as a night scene with moving traffic and crowds of people and then, using the aerial image, simulated 'neon' lighting gave the name of the programme and the effect was convincing. The cost of constructing a sign, which was considered, was avoided.

Video versus film
Advances in video animation, the increasing use of computers and their obvious compatibility with the whole of television production have taken much of the impetus away from film animation; but film has the advantages of lower cost for capital equipment and operation, and can be managed as an in-house service for large areas of television graphic output.

Many of the examples of creative graphic titles illustrated in this book were shot on rostrum cameras; from the 'Anatomy of a Murder' by Saul Bass in the mid-fifties to some of the best titles shown on our television screens everyday some thirty years later.

The film system is still a useful servant. In the hands of those who chose to continue to work in this medium it will contribute much more.

Section 6b
Model animation

Television graphic designers have used models on many occasions in their search to produce animation and special effects for titles as well as programme material. Filming, and more recently recording on video, a three-dimensional object, or model, gives the ability to create movement, to shoot the subject from many different angles, to change the lighting effects, and to play tricks with different scales. All this can be achieved without the laborious use of drawn and painted cels, or the usually higher cost of computer-aided animation.

Some model animation work approaches the surreal quality of the much admired fully computer-aided graphics, (compare the 'County Hall' title on page 112, while the title for 'Tomorrow's World' filmed a succession of moving objects with special effects. (Page 113).

Some model animation work uses film cameras which can shoot one frame at a time. To gain smooth and precise movements the camera is mounted on a dolly, usually motorised, and fixed to a rig which is very solidly built and can be moved by very small increments.

A model animation studio requires a big area so that large-scale models, sets, or even real objects can be viewed from considerable distance and from the widest possible angles. Varied and specialised lighting is essential.

Once the model is commissioned, (in some cases when there is not much detail this can be done in the graphic department) the cost of filming is usually well within the limits of the budgets and the time often is shorter than other forms of animation. When the model is very complicated, or for filming has many articulated parts, the skills of the professional model makers are brought in.

There are many companies who provide these basic necessities for model animation and a few who offer highly-specialised equipment and facilities for this type of work for the television and film industry. High-speed cameras can record fast action at hundreds of frames per second.

Motion-control model animations
While a mounted film camera can produce a wide variety of model animation a more precise system of motion-control has been devised. By building a camera rig which can be moved on up to 11 axes, each moving at increments of 1000th of an inch, with computer-control, much more precision can be obtained. A rig can also be made to move a model with the same degree of computer-control.

One company which has done this is The Moving Picture Company. The picture on page 114 shows the camera rig and the overhead tracks in the studio. Importantly, a video or a film camera can be used as required. Video has the advantage of speed. 'The main reason we did the job on the video rostrum was time. It was shot on Good Friday when the labs were shut, edited over the weekend and ready on Monday.' said the operator at one company. The disadvantages of video, according to another production house, are that the resolution is well below film standard and 'The models have to be lit until they nearly melt before there is enough light for a video camera.'

With a 'snorkel' on the lens either a film or video camera can be made to 'fly' around a large scale model up to 10 feet by 10 feet, or as small as an egg cup, to extremely fine tolerances at any apparent speed.

Whatever the moves and the exposures the entire movement within any sequence can be stored in the memory and then relayed over again with perfect repeatability, or adjusted to make any amendments found necessary.

Section 6c
Video rostrum systems

The search for an alternative way of preparing animation has been progressing for a long time – in spite of the great improvements in the use of the rostrum film camera.

Recording multiple images on video has two immediately clear advantages. The results of video work can be viewed by the cameraman and the designer as the work proceeds and corrections made. This is an enormous advantage when compared with the film system where no check can be made until the film is processed. Secondly filmed animated titles and inserts are of a noticeably different resolution and picture quality to the video tape material with which they are combined. They also need a telecine machine for replay.

Early ways of using a video camera for single frame animation were simple. The camera was fixed to an overhead rig where it could be moved in small increments on x and y coordinates above the artwork and the exposures were recorded on videotape and replayed as required. These units were known as 'four posters' and an early unit was made by Evershed Power Optics.

Later stages had a fixed video camera, with zoom, and a conventional film rostrum bench to manipulate the artwork.

The major breakthrough came with the use of video stills stores using large capacity disc packs. These allow random access to the images and the frames can be edited digitally in endless computations. Manufactured units using video animation have not been widely introduced into television companies so far. Prior to the use of disc storage the attempts to mix images required the use of two video tape machines. This was expensive and clumsy.

The BBC video rostrum
In 1983 a detailed description of the BBC large scale effort to build a video rostrum camera and digital storage system was given in a paper at the BKSTS '83 Conference*. The work has resulted in video animation system devoted to providing the graphic designers at the BBC with a what they described as 'a flexible

* Generation of animated sequences for television using an electronic rostrum camera and digital storage system.' Kirby, Devereux and Wolfe/BBC Reproduced in the Journal of the Royal Television Society, March 1984.

means of creating animation using television techniques and the facility for instant appraisal of their work.'

No smaller organisation could possibly have sustained the expense and duration of this experiment, however successful.

The video opticals for 'The Life and Loves of a She-Devil' are an example, page 00, of work produced on the BBC system.

The one unit installed so far is not able to process the vast quantity of material handled by the many film rostrum cameras but the reaction of the graphic designers is very enthusiastic and the pioneering research and development is bound to lead the way to further use of electronic animation.

Section 6d
Colour separation overlay

Using more than one television camera to make a combination picture for transmission is a very widely applied technique involving graphic design at many levels. Its most creative applications allow the set designers and graphic designers to work very closely together in a way many people believe the new electronic equipment will lead us in future. The scenery for a complete studio can be substituted by a small illustration, or model, or originate from a digital paint system.

At the BBC the generic term 'CSO' (colour separation overlay) is used; while ITV use the trade names 'Chromakey' and 'Ultimatte'. In most cases a studio is dressed entirely in one colour, usually blue, (this is less likely to be seen in flesh tints,) and one camera views the action. Another camera records a drawn caption, or photograph, and these need only be a few inches in dimension. Small models or real objects have also been included.

The illustration and preparation of these backgrounds are the responsibility of the graphic designers and the challenge to make them as inventive as possible has resulted in some very interesting results.

Cutting costs
The purpose is to combine actors in a studio with minimum, or no scenery, or combine a presenter on a news programme, with graphic or photographic backgrounds. In the first CSO experiments the effect was often very obvious. 'Tearing' occurred – that is rough edges particularly around moving figures, and there were no

cast shadows.

This limited the effective use of CSO to fantasy effects and unreal situations. Later more and more realism has been possible. But recently whole programmes have been based on the technique and two examples are, a series made by the BBC based on the Second World War newspaper strip-cartoon character 'Jane', and Thames Television's production of Berlioz's 'L'Enfance du Christ'. In the 'Jane' production the aim was to make all the drawings, which were drawn on 'Paintbox', look as much like the newspaper originals as possible. To help this the live characters were shot in black-and-white video and then electronically tinted to blend them into the illustrations. The whole series involved hundreds of drawings and the entire production was planned for graphic and CSO production.

Design co-operation
'L'Enfance du Christ' was designed with the close co-operation of the graphic designer, Barry O'Riordan and the production, or set designer, Peter le Page. Berlioz had conceived the work as an oratorio – not as a fully staged opera – and suggested a series of tableaux 'in the manner of the illuminated missals'. Page and O'Riordan, and their Producer John Wood, decided to use a variety of paintings by the Victorian orientalists from the late 19th century which is very rich in representations of the Middle East and Holy Land. 'Ultimatte' was then an excellent device to recreate, with the musical passages, the actors and the landscape paintings, composites which would have been impossible using the scale of the normal television studio settings and painted scenery.

The extensive subtitling and captioning was produced on an Aston III generator in a typeface especially designed for the production and digitized on to a floppy disc. The lettering was colour-matched to each scene using the joystick colour facility on the Aston.

In daily use the captions and illustrations prepared by graphic designers keyed into news bulletins behind the presenters are the most familiar use of CSO, but more creative uses have been employed in almost every area of programme making.

Childrens' programmes have used the system where budgets are low. Many changes of background can be made if the scripts require them and fantasy and dream sequences became relatively simple.

Section 6e
Digital effects and still-stores

Manipulation of still and moving video images has become highly developed in television production but because of the instantaneous, and at times, almost random, nature of these 'stock effects' they have not always sat happily with planned graphic design.

Scott Miller of CBS New York: 'In the past special effects devices have always meant trouble for television graphic designers. Early on, video switcher wipes created graphic, often ghastly patterned transitions between pictures in programmes and commercials . . . A host of video tricks and gimmicks have undermined designers efforts for more than 20 years.'

Uncontrolled effects
Broadly this area of picture movement has been controlled by directors, producers and technicians without much time, or thought being spent on each sequence. That was digital effect's virtue – they often by-passed the whole process of design and the graphic designers! The results often *appeared* to be the responsibility of graphic designers as these 'roller-coaster' wavy-line images and wildly spinning shapes cavorted on screen.

Every television studio has an adjacent Production Gallery to control the recording, or transmission, of video material. Galleries have always had the equipment to affect the pictures produced from simply fading-in and out, to mixing the picture sources, and a wide range of 'wipes' and image distortion – all controlled from consoles. These now incorporate banks of devices and allow pictures to be juggled by Vision Mixers in many ways – all preprogrammed and operated by simple switches or joysticks.

At this level graphic designers have little control over presentation where programme making has so much emphasis on immediacy and constant speed.

Attempts to hold-up the process by checking and revising design decisions and pre-programmed effects has proved impractical so close to transmission.

Editing suites
VTR (Video Tape Recording) Edit Suites are the centres within television stations where, among other editing activities, digital

effects machines are available for graphic designers to work with technicians to compile video sequences to their designs or storyboards. They can transfer film and slides to video, insert character generated lettering, edit video frame-by-frame and then add any of the effects which the particular Edit Suite provides. The age of the complete digital video studio was proclaimed by a number of manufacturers at the 1986 IBC (International Broadcast Convention) Exhibition at Brighton and it is vital that graphic designers keep control of graphic effects in the new devices.

Digital designs

Graphic designers can and do design for digital effects machines – Quantel's Mirage, the Ampex ADO (Ampex Digital Optics), and many others have been widely used in many title and programme inserts planned and produced away from the immediacy of the gallery. The aim is to use the picture movement, or distortion, for a reason – not let the clever gimmick appear just for its own sake.

Graphic designers prefer computer and electronic aids they can 'drive' themselves, (like digital paint systems), where the machine's ability does not overpower the style, and the resulting image does not have the heavy-hand of one particular effect: they also prefer those where there is time to plan and approve the design work before transmission. As Colin Cheesman said, when he was Head of the Graphic Design at the BBC, 'Electronic devices provide instant pictures, which means there is a danger of expecting instant design. The designer could find himself more concerned with speed to the detriment of imaginative solutions.'

Outside facility houses now offer a mass of digital equipment for graphic design video production and animation and they all possess their own recipes for the mix of hardware. The combinations are now exotic in their range and complexity. Perhaps they will reduce and simplify in the future. One company lists: Quantel's Harry, with access to Paintbox; Encoré, Bosch FGS4000 and Telecine. Another offers: Ampex ADO; two channel Quantel; Abekas A64; Telecine; Paintbox and Ampex Cubicomp Picture Maker.

Trying to select the most efficient and suitable venue is similar to choosing a car when there are seven models in the range and options within each of those!

Graphic designers must view the various showreels, discover just what each company can do, and use the 'grapevine' to learn from other graphic designers the cost and speed at which work can be carried out.

Storing still frames

Stills stores are the key factor in the advance of electronic graphics. There was little chance of any progress in manipulating electronic graphic images until they could be recorded instantaneously and held in a stack, or library, in very large quantities for equally instant replay. They are known as Electronic Stills Stores (ESS).

Electronic technology now enables full-colour pictures from any source to be 'digitized' into binary numbers and stored on to discs. An early Winchester disc had a storage capacity of 165 megabytes – the equivalent of about 160 frames. Any stored picture can be coded and re-called as required.

A digital paint system can be seen as an essential tool to a stills library, and not as a graphic designer's creative outlet. With a paint system, pictures in the store can be cleaned, re-touched, reversed and altered in many different ways.

Before the age of the stills store multiple pictures were held either as separate pieces of artwork, usually called 'captions', or as large collections of 35 mm slides to be transmitted through a telecine slide-scanner. The stills store records, holds, and re-plays all in one device, thus eliminating three separate processes. Photographing the original, storing in a carousel, and then accessing through the scanner.

Banks of stills stores increase capacity even further and these are termed 'Digital Library Systems' (DLS).

Stores, also known as frame-buffers, hold in computer memory pixel data of separate frames for raster presentation and were essential to the development of computer-aided animation.

Section 6f   Live-action

Hand drawn and computer animation have been shown to be slow and costly. However, a simpler, and often economical, way of producing movement for television sequences lies in the exploitation of live-action. This can vary from the decision to stage carefully planned live-action and

shoot, almost frame-by-frame, from a properly designed storyboard to a compilation of material selected from previously videoed, or film footage, with lettering and effects attempting to bind the sequence together.

Library material

Filmed library material, either archive or contemporary, can be the basis of contributing part of an animated sequence. Here the important point is to take great trouble to research the best and most striking examples from the range available and to ensure it is well presented, or treated in an appropriate way. Film can be reframed on an aerial image of a rostrum camera and now it is relatively simple to adjust video material in almost any way required frame-by-frame using a digital effects machine.

Location and studio shoots

Many ideas interpreted in live-action lead to location shooting, or setting-up scenery, or objects in a studio. For a Bank Holiday promotion sequence a traditional fairground roundabout was hired, assembled in Pinewood Studios, and set against a large cyclorama with a projected sky and the whole thing set in motion.

The problem with live-action comes with ensuring the film, or video, matches the visualised idea. The vagaries of weather conditions and the control of the action can be very nerve-racking and on occasions extremely difficult to repeat.

Classic applications

A memorable use of live-action was Pauline Carter's title for the BBC sports programme 'Match of the Day' (1975). Her bold idea used mass display picture-making. The pictures included players, a referee, cups and portrait of presenter Jimmy Hill and they were all composed by people in the football stadium holding coloured panels above their heads.

TVam made a spirited start when they first transmitted in 1982 using live-action to good purpose in Ethan Ames cheerful 'Good Morning Britain!' spelt out with considerable daring by freefall parachutists, pigeons, naval ratings and a friendly crowd at dawn on Bristol Downs.

Live-action can involve machines, animals and natural effects on land, sea or air – but whatever is used it should not be a substitute for creative design solutions.

Section 6g
Digital paint systems

Paint systems are packages with user friendly, interactive, palette-driven software, which can be learned fairly quickly. The hardware generally consists of a computer, frame buffer, cathode-ray (CRT) tube monitor, digitizing tablet and pen, terminal, disc drive and various input and output devices like film recorders, and video input or output. Pictures are created in real-time using instructions input with an electronic pen on a digitizing tablet. The image appears on the CRT. The palette-driven software consists of a menu of options (commands) chosen with the pen and tablet. *

In the early eighties when the electronic revolution was gaining momentum Ampex, the American manufacturer produced a sales video cassette to promote their digital paint system called AVA. (Ampex Video Art.) This new unit was very advanced for the period but the price at that time did seem high. (Around £120,000. Quantel's 'Paintbox' DPB 7001 is now approximately £70,000 six years later.) The sales angle of an Ampex promotion cassette highlighted the lack of change in the television art department, and the commentary said:

'In 1956 a video tape recorder weighed over half a ton and was nearly the size of a car. Today it can weigh as little as 50lbs and be carried in a car beside you.

Thirty years ago what little video tape editing was carried out was done with a razor and the eyeball. Today it is push button simple and accurate to a single frame!

Thirty years ago the art department looked like this. . .

(A black-and-white photograph of a designer at a drawingboard – surrounded with brushes, paper, pencils, glue and general studio clutter appeared.)

Then came the punch line . . .

Today the art department looks like this . . .!

The picture of the same simple, cluttered, studio appeared unchanged except it was in colour!

The point was made. Technology had by-passed the graphic designer in television for thirty years.

*Definition by Darcy Gerberg SIGGRAPH Technical paper 1983.

Computer-aids arrive

In most countries change has now occured. Character generators were among the first to arrive; digital paint systems, computer-controlled film and video rostrum units, and computer-aids for animation have followed in all but the smallest design groups.

With many others associated with television production – engineers, graphic designers, and technical staff – I first saw a demonstration of a digital paint system about seven years ago.

The impact, like any real 'first' was very strong indeed. Like hearing Edison's recorded voice reverberate from the first phonograph, or seeing Baird's early television demonstrations.

A 'pen' in the hand of an artist moved uncannily across a continually blank white tablet while his eyes concentrated on the video monitor ahead. Miraculously the screen rapidly filled with lines and shapes, as on a sketch pad, later to be removed, changed in colour, and over-layed until a final picture emerged. From abstract pattern to considerable realism, the range was there.

Too much emphasis was placed in these early demonstrations on 'fine art' pictures which resulted in the epithet 'The most expensive piece of paper in the world' being unjustly applied. Not only the individual freehand style of the artist was translated into the medium of video but these systems provided the facilities of all forms of geometric drawing aids. Perfect circles and ellipses take only seconds to construct. Horizontal and vertical lines can make grids and aid all manner of pictorial and diagram construction. Even complex cut-out shapes can be enlarged, or reduced, and then moved via a 'stencil' to any part of the screen.

Artwork of very high-finish, simulating every type of conventional drawing and painting can be prepared and displayed extremely fast. Airbrushing, transparent watercolour, pen, pencil, chalk and even the impasto of oil paint can be represented.

Speed of operation is the most important benefit. It would take at least two hours to produce a well-finished piece of artwork of the complexity of the early US flag shown on page 128. On the Quantel 'Paintbox' the job was done in about 30 minutes!

A single image can be automatically repeated endless number of times at any size and the range of colours available in many of these devices with 24 bits per pixel

is around 16 million colours.

Recent additional software has added 'perspective' improvements and 'blurr' to soften parts of the picture.

Rapid progress

The early devices were improved very quickly. At one point there was only enough memory to digitise black-and-white photographs through a monochrome video input camera. Very limiting indeed in colour television transmission. Shortly after full-colour stills of any image – paintings, photographs and even real objects – could be frozen on the monitor, digitised, and immediately incorporated in any design.

Within two to three years a prototype of their digital paint system was shown to me by Quantel and a video camera recorded the view from an office window. The houses and garden were 'frozen' on-screen as a still frame and the demonstrator instantly picked-up the exact colour of one tiny patch of sky and 're-touched', with the airbrush mode, to remove the aerials on the roof tops. This was done so perfectly we were impelled to turn our heads and look outside the window to see if they had really disappeared!

There are now many competing systems of various prices and power. Digital paint units have been used for preparing illustrations for colour separation overlay, and for a great deal of computer animation but their application to other areas within television is immense. Set designers have already begun to use them (Particularly at the BBC) and aids to costume design and make-up are foreseen.

Their fluency in assembling photographs, drawings and engravings – almost anything – to make collages and adapt images very rapidly has proved very effective in the constant output of on-screen promotion slides. Channel 4 in London has made excellent use of 'Paintbox' and the lively and imaginative results of this work are shown on page 69.

A high-definition (1200 line version) has already been developed, and outside the television industry digital paint systems are expanding into design for printed graphic design in publishing and advertising, architecture and industrial design.

The concept of the graphic video workshop

Running parallel with the use of these paint systems the use of stills-stores in television has developed just as dramatically.

Hundreds and even thousands, of individual stills each containing digital information for the entire transmittable picture, containing about 300,000 pixels per frame can be stored and instantly recalled for transmission.

Still-stores and paint systems are equally useful and equally dependent on each other. Their power to create, store and recall artwork, made the development of electronic graphic workshops a reality, particularly important to news programmes where the whole working environment depends on speed.

An early example was the marriage at the CBS graphic design group under Ned Steinberg in New York where electronic artwork, produced at high-speed on the AVA, was linked to an Ampex electronic stills-store, (ESS) as early as 1980. The revolution in graphic presentation for television news programmes had begun. Since then electronic graphics have spread to almost every country in the world and involved graphic designers in news presentations on a greater scale than was previously found possible.

Hard-copy off-screen photography
Recording video images off-screen as colour prints or transparencies was for a long time very unsatisfactory. Using an ordinary stills camera reproduces the line resolution which makes further copy by any system less attractive. With the introduction of computer generated graphics, where the only image is the picture displayed on the monitor, the need for something other than video became obvious.

Cameras have been developed using various ways to enhance the off-screen picture. The principle they apply is to separate the red, green and blue (RGB signals) and photograph these one at a time and then realign to overprint and conceal the coarse line effect.

Among cameras which do this are the Honeywell, Dunn, Dicomed and Matrix and they all must interface through a stills-store, or frame buffer. Most can produce copies in a variety of formats in either colour transparency, Polaroid, or colour negative material.

Centre stage
In the 1986 Royal Television Society Schoenberg Memorial Lecture Richard Taylor, the Managing Director of Quantel said, 'As Paintbox appeared on the scene,

there was a degree of concern amongst graphic designers that their livelihoods were in jeopardy due to the much increased productivity possible with such devices. I am pleased to be able to report that far from there being long queues of unemployed designers in the industry, in fact there are, today more employed than ever before, thanks to the increasing involvement of graphics with TV programmes. It is not an exaggeration to say that electronic graphics have moved graphic designers from the wings to centre stage.'
Has it taken nearly thirty years to discover that pictures are important to television?

Section 6h
Computer-aided animation

Making better pictures
Mathematics and art have collided on a number of occasions in the past. The Greeks used mathematical proportion and geometry in the construction of their architecture and the Golden Mean ratio applied science rather than intuition to picture-making and building.

One of Paolo Uccello's experimental drawings of 1450 used a chalice to depict the perspective of a solid object. Over 400 years later one of the most important pioneers of computer-aided animation, the American James Blinn, chose a chalice to demonstrate the way in which pixels on a video screen can, when controlled by digital computers, form any object and create backgrounds. Uccello's drawing, looks like wire-frame computer output, and even forecasts the 'hidden-line' concept — an important advance in the early development of computer-aided animation.

A new dimension
Dr John Vince, who was until recently the head of computer graphic studies at Middlesex Polytechnic, expressed this view: 'I doubt whether the pioneers of computers anticipated the impact computer technology would have upon the graphic arts. Who could have predicted in 1950 that within the space of three decades it would be possible to display three-dimensional, coloured, animated images in real time?

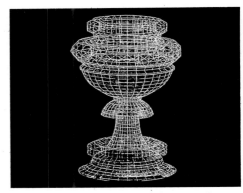

Science and art: Uccello's 'wireframe' drawing from the 15th century and below a mass of numbers produce incredible detail in a 20th century moving realistic image.
'London Plus'/BBC/UK
Graphic designer/Lesley Hope-Stone
Computer Animation/Electric Image

The role of computers in animation
A succinct and understandable description of the role of computer animation and its application to graphic design has been made by Nadia Magnenat-Thalmann and Daniel Thalmann of the University of Montreal. I quote from their brief introduction to 'An Indexed Bibliography on Computer Animation'.

'Although the computer plays an ever increasing role in animation, the term 'Computer animation' is imprecise and sometimes can be misleading, since the computer can play a variety of different roles. A popular and simple way of classifying animation systems is to distinguish between *computer-assisted* and *modeled animation*.

Computer-assisted animation consists mainly of assisting conventional animation with a computer. This type of computer animation is carried out mainly in two dimensions. The computer can be used to input drawings to produce in-betweens, to specify the motion of an object along a path to colour drawings and create backgrounds, to synchronize motion with sound, and to initiate the recording of a sequence film.

In *modeled animation*, the computer becomes more than an aid – it plays a basic role in the creation of a three-dimensional world. This type of computer animation involves three main activities: object modeling, motion specification and synchronization, and image rendering'.

Computer-assisted animation
Inputting the drawings
Linking computers to animation was a 'dream' which would reduce the extremely slow and difficult task of preparing hundreds of cels that have to be drawn, traced and then hand-painted on the reverse side in conventional 'cartoon' animation.

There are computerised aids which very directly copy this old system. Instead of the animator's drawing being hand-copied on to acetate, the outline is digitised into a frame-buffer. It can then be re-called on to a monitor where area-by-area, it is quickly coloured electronically by a digital paint system similar to Paintbox. Each 'cel', or frame, is then stored to be assembled in any order to produce the animated effects required.

It is important to remember the above aid does not reduce the animator's work in drawing every 'in-between'.

The producers of large amounts of drawn animation and feature-length cartoons have applied these labour saving techniques, but they are not generally available to graphic designers working in relatively small units within broadcast television around the world.

A computer can be asked to accept a drawing as a starting point, for example the outline of a butterfly, and a drawing of, say, a seahorse as an end point. Then an instruction to draw as many in-betweens as the designer requires is given to fit the length of the animation. The result will be multiple drawings – each slightly different from the last – to make the required metamorphosis. This was one of the techniques available on the 'Picaso' system at Middlesex Polytechnic in the early 80s and harnessed to a high-speed line plotter it proved useful for a brief period. An application is shown on page 55.

Using a computer to specify motion along a path
The computer can be instructed to aid the animator not only to make a metamorphosis but to plot the motion in space of apparently three-dimensionals objects.

The drawing can then be used as the first basis for producing hand-painted cels to give the animation form and colour. 'Music in Time', designed by Robinson Lambie-Nairn where the metronome, the complex drawing of the Albert Hall, and the other objects flying through the air were generated on the same Middlesex Polytechnic computer in Dr John Vince's department in 1981. (page 132).

Modeled animation and computers
There are many examples of this role of the computer in the illustrations at the back of this book, under Section 6h, and it is the result of the advances in this field which bring to the viewers mind the words 'Oh, that's computer graphics!' because they present a complete environment where objects look fully-modeled, react to light and produced shadows and reflections in a quite uncanny and realistic manner.

An attempt to simplify the way this is achieved is made by diagrams and notes on page 130 as 'graphically' as possible. Graham Kern's storyboard and choreographic outlines for the 'Paul Daniel's Show' (Page 58) demonstrates that the working out of ideas has not changed from the days of filmed graphics.

In *modeled animation* the computer takes over and creates a complete three-dimensional environment. To do this it uses object modeling, motion control and image rendering. Within this work computers have been programmed to produce the reflections and shadows of moving objects, as well as provide surface textures and appearances from highly-reflective metal to transparent perspex. It is this element which has such a dramatic effect on the observer who sees a creation of the real world made from a vast list of mathematical calculations. Building-up the separate frames of animation can take either hours or minutes of computer time. This depends on the power of the computer and the amount of detail, or information, each contains. This process, called 'rendering', is still incredibly fast for what is achieved and often completely beyond the hand-animation system to even attempt.

Graphic designers have been able to plan storyboards for the newly developing system and then present them to the rapidly expanding facility houses and computer animation companies which have appeared throughout many countries in the past few years.

In America, where the state of the art was pushed so far and so fast, the first impetus was from the US space programme – then through the engineering departments of many of the Universities. Massachusetts Institute of Technology, the New York Institute of Technology and Columbus, Ohio were among the leaders. America does not have the restraints which separate commerce and academic work, and these colleges were among the first to sell their new-found skills to the television and film industries.

Spreading the picture
'Art the most significant bit' A lapel badge, with this neat pun (a bit is the smallest unit of digital memory, a contraction of the words 'binary digit'), was worn by some brave people at the annual *SIGGRAPH conference in Minneapolis in 1984.

Why brave? SIGGRAPH is the world's largest gathering of the *computer programmers* and *engineers* working in the graphic arts industry. They, and not the artists and designers, dominate this vast international display. The badge was a reminder that no amount of computer hardware, or millions of dollars of software, can make a poorly conceived idea into a good one.

When there is a confusing amount of jargon, and equipment, plus a very rapid change, in film and television design this emphasis on the simple basic principle of *design* is very reassuring. The advances and the excitement of the past years in computer graphics has been revealed and monitored by a number of international conferences, and the exhibitions of equipment. SIGGRAPH celebrates its fourteenth year in 1987 and this vast event, which has been influential in spreading knowledge, both academic and commercial, about all aspects of computer graphics will move to Anaheim. Parigraph spreads the word, or the picture, in Europe every year with a spring exhibition in Paris, and Online have a Computer Graphics Conference in London in the autumn.

*SIGGRAPH Special Interest Group in Computer Graphics.

How friendly is user-friendly?

The electrical basis of the computer is the ability to calculate and execute a given programme of instructions of immense complexity at incredible speeds. When you consider that a single frame of full-colour picture on a video monitor screen contains about 300,000 pixels, made up of the red, green, and blue signals and these frames are cycling at 25 frames per second, the amount of power required to control pictures as complex as those in fully modeled computer animation is almost beyond comprehension. 'User-friendly' is a reassuring term which prevents all those graphic designers who wish to use computers to present their design solutions from losing heart.

They do not have to write programs or explore the rarified mathematical problems of Cartesian co-ordinates, or understand algorithms and the contribution of Bui Tuong-Phong who in 1973 described 'the use of interpolated vector normals rather than light intensities' and created Phong shading.

Human modeling and computer animation

Computer-aided sequences have performed well when they handle geometric and abstract forms. The laws of light and reflected light and the control of solid-appearing polyhedra is well inside their ability. Relatively anamorphic objects, the human form and face, and animals and plant life are still a difficult challenge.

A few landmarks in figure animation have already been reached; among them – Tony Petries 'Piano player', Robert Abel's 'Sexy Robot', Hitoshi Nishimura's and Takashi Fukumoto's 'Bio-Sensor' – a dramatic skeletal animation of a panther in motion, and the 'Snoot and Muttly' exercise by Cranston Csuri. But these have all been done at normal commercial restraints.

A new aspect has arisen and recently Martin Lambie-Nairn yoked the skills of an 'old-style' animator to a facilities company who were producing modeled computer work for him. While the job was in progress a cel animator was employed to develop the human expression and movement to a higher standard than the computer animator/programmer could conceive.

The creative idea of making a human-being imitate the *limitations* of computer images through the 'computer-friendly' Max Headroom is a lovely illustration of the predicament.

The cost of the computer

The time and cost of developing fully-modeled computer animation has been very high indeed but it's rapid acceptance has reduced this. When Robinson Lambie-Nairn were producing their animated indent for Channel 4 just over six years ago they had to go to America to get the job done. Now there are at least a dozen companies in London alone who could tackle the job as routine.

Section 6i
Lettering for television

Although lettering has been a basic need in every programme the contribution of television to the art of typography has, until recently, been poor,

Through the efforts of some television engineers and graphic designers, (among them John Aston, Graphic Design Manager at the BBC,) the manufacturers of electronic character generators have slowly brought television into the twentieth century. Typestyles, spacing and compilation and composition facilities are now available to meet the requirements of the discerning designer.

At a conference held about five years ago, speaking on behalf of television graphic designers, John Aston said, 'Our objective is to ensure that at each stage of development new automated systems aid, rather than erode, traditional design values and craft skills'.

Pre-electronic lettering

The earliest way of providing lettering for television now seems very primitive. Most captions were hand-painted on black caption cards and placed in front of a studio camera.

Low resolution, which affected all forms of television graphic work, was particularly noticeable on still rather than moving images. Type on-screen therefore suffered a great deal from the technical limitations. Type had to be large and the characters avoided any thin strokes or fine serifs until some improvements came with higher line definition and improved receivers. Bold typefaces – Announce Grotesque, Grotesque No. 9, and Egyptian Expanded were among the most prevalent for many years.

The system of setting and proofing type that television made its own, was the hot-foil hand-press using type founders moveable metal type. Almost universal was the 'Masseeley' press made by a company called Masson-Seeley – it had been widely used for printing price tickets and small display notices in shop windows.

A single impression could be made from the type heated by an electric element. The virtues were: no ink was required hence the impression did not need to dry and a perfectly solid white letterform could be made on a black background. This was ideal for television of the period and over-laying lettering on most pictorial images as bottom-of-frame captions and simple titling was reasonably satisfactory. Printing type onto clear cel, or onto artwork and illustrations, for rostrum film work was an additional everyday use. But the whole process was very slow. Type was hand-set character by character then: Locked into a chase and placed on the platen: Proofed on to caption card, or on many feet of paper for a roller: The caption, or roller, then had to be lit and viewed by a television camera for transmission, or photographed as a 35mm slide for transmission through a slide scanner.

This took hours rather than minutes for even small amounts of copy. With a careful operator the typographic standards of letter and word spacing could be very high.

Letraset and other forms of transfer lettering were eagerly seized upon in the late 1950s. These supplied a wider range of faces and enabled graphic designers and assistants to prepare more artwork, control the paste-up and letterspacing and be less dependent on the Masseeley printers.

In the sixties photosetting, mainly developed for the printing industry, was a useful addition to the production of television lettering. It's main advantages were the ease of setting at varied sizes, obtaining lettering white-on-black in one process, and greater variety of faces by reference to almost any typefoundry in the world.

The arrival of the electronic character generator

There are now many different character generators throughout the world but they all operate from a keyboard similar to a typewriter. At their simplest they produce a letter, or numeral, on-screen for every key depressed and most now store a library of typefaces which can be changed in style and size at the touch of a key. The faces are generally stored on floppy discs and now alphabets can be digitized into the computer, through a font-compose device, in a matter of a few hours for any new design.

The first character generators did not improve the typographic standards and a model which arrived at the BBC in 1969 called 'Anchor' (Alpha-Numeric Character Generator) had poor letter forms without proportional letter spacing. Every character was, as in most typewriters, on the same width body.

The advance was important technically – the generator could be used in live programmes – but the aesthetic effect was disasterous.

The next generator to arrive on the scene was the 'Ryley'. This gave better resolution and the letter forms were based on Helvetica with nine variations of width and weight as well as proportional spacing.

Since then the range of generators has expanded. Aston III and Aston IV, Chyron, Telemation, Dubner, Vidifont and Cypher are just a few of the makes on the market.

They supply hundreds of faces, and Cypher takes the character generator into the realm of computer animated typographics because it allows words and even individual letters to be manipulated in almost every conceivable movement.

As the characters are made up of only a few pixels the letter forms tend to be very jagged indeed. This defect has largely been overcome by putting in more information to each character by adding pixels of varying grey scales to soften and enhance their appearance. This process is called anti-aliasing. The Aston IV software anti-aliases all type and the improvement in letter definition is very marked.

In sports events and news programmes character generators have a very special part to play, for the presentation of results, scores, and the late information can be virtually instantaneous.

The speed of setting, the libraries of typefaces now available, the clarity of presentation and the storage of pages of composed information to be recalled at the touch of a key, have made character generators indispensible to the graphic designers and technical operations in almost every television station in the world in less than a decade of research and development.

The situation for graphic designers is broadly represented throughout this text as working for large, national, or government licensed, networks, and this system has pertained in most countries for five decades.

The explosion – so many anticipate will change the structure of television broadcasting throughout the world – is only just about to be detonated.

Direct broadcasting by satellite (DBS) and transmission via cable are going to change the ownership and operation of networks, as well as the habits of viewers, in Europe, America and the rest of the world. The increase in privatisation is likely to lead to many more smaller broadcasting and production units plus an expansion of the number of independent television programme makers.

This proliferation will increase competition and the number of graphic designers required.

Will the changes improve or weaken the standards of design and graphic presentation? Will the pressure to maintain large audiences, and profits, result in the loss of committment to what has been achieved to date?

Design Management
A key factor will always be in the hands of those who manage television companies. They need to see design as a positive way of gaining audiences, making profits and providing new and better services – similar to the current design revolution in the retail business in our High Streets. Not as an unavoidable adjunct to television programme-making.

Future designers
There is a clear enthusiasm to work in television. People can be drawn by a misplaced idea of the 'glamour' of the business. One 'unknown graphic design candidate' wrote in his job application: 'I feel television provides a unique opportunity to demonstrate graphics in unlimited form, and very rewarding when seen to be communicating to the public'.

Art colleges would not be wise to concentrate too much attention on specialised training for large numbers of students directed towards television graphics.

Those who discover a strong interest and aptitude for animation and television will surface. Film and video equipment is being

introduced into many colleges – often those where there has been a strong tradition of animation work and rostrum cameras.

Computer animation equipment is very expensive and so are electronic drawing devices if they are to match broadcast standards. Low price systems, like Pluto, have been installed to good effect in a number of Colleges. A few regional centres in the UK were fortunate in being selected under a Department of Trade and Industry for possession of Quantel's Paintbox. These were Middlesex Polytechnic for London, a unit went to the Midlands and another to Scotland.

Creative aspects are paramount and high standards can be achieved without equipment. Even the professional designer, like the student, is still faced with the imperative of thinking of an idea and then conveying this through a storyboard. The annual Quantel Bursary for television design has stimulated interest and proved that large amounts of hardware are not essential to creative thought.

'The Changing Image'
New technology has brought new problems and Geoff Crook wrote recently in the book which accompanied his unique exhibition at the Central School of Art and Design in 1986: 'The Changing Image'

'The aesthetic challenge of graphic design for television has not been driven away with the easy solutions which the influx of new technology present. Indeed the problems of design have arguably been complicated and exacerbated as a bewildering array of special effects facilities constantly pressure the designer into the temptations of movement and surface gloss as a solution for every problem.'

Computer-aids are largely responsible for the view that there is not as much genuine inventiveness as there should be. A strong expression of this came from Paul Brown, the Head of the National Centre for Computer Aided Animation and Design. 'Look at television design anywhere in the world and they are all copies of the same thing. The problem is stagnation. There is no innovative work, no one is using the computer as a new medium.'

Only by continually responding in an imaginative and original way to the task in hand can the process of graphic design maintain the purpose of explaining, and revealing, the message of the programmes.

Graphic presentation over forty years

BBC Television archives date this film animation as 'circa 1955'. The date coincides with the arrival of their first formally appointed graphic designer. There is no credit to this piece which so strongly resembles the cinema news titles of the 1930s and the RKO trademark.

The image of the Alexandra Palace mast, site of the first regular television transmission service in the world, has lingered in the imagination because it has recently been used as a pastiche for the titles of the programme 'We Love TV'. See page 102.

Station identity

No matter what efforts are made to exploit the moving image a single 'still frame' stays in the mind and is required to portray in print, stationery, on vehicles and other media the face of the company or network.

Inevitably artwork and slides for the transmission of these symbols have slowly given way to new technology and some of these are shown on the opposite page.

Graphic designers have usually been involved in the design of television station clocks. In the mid-1950s Leslie Mitchell, the Presentation Controller at Associated Rediffusion, commissioned this metal design suggestive of paper sculpture. A television camera in Telecine viewed the mechanical clock, and it was later replaced with a

photograph of the surround. The original is now in the Science Museum in South Kensington. The numerals were Latin Wide — a favourite typeface of television designers at that period, and the radiating sun (top centre) was Rediffusion's symbol.

Granada's on-screen animated ident and symbol has, like many other television companies, been subject to changes over the years. Three stages are shown here — from a black-and-white animation to a recent fully computer-aided version. The symbol of the now extinct ABC Television company served the UK company for fifteen years.

Abram Games, a respected figure in many design fields in the 1950s, was commissioned to design the on-screen identity for the BBC. A frame from a regional version from the BBC North of England is shown here. The use of well-known and established designers outside the television system was unusual then and still is rare.

BBC2 Computer ident/1979
This symbol was designed by Oliver Elmes to be animated by the BBC Computer Graphic Workshop. Two four-second versions could be played out in real time by a solid state 'black box' designed and built by the BBC Engineers. Previously idents were static or run from film or video tape. The new system always presents a first generation video image.

William Golden's 'eye' has served CBS very well indeed. What came from his drawing-board in 1951 has appeared in every aspect of CBS operations in most countries over the past thirty-five years.

BBC1 Computer animation/1983
An old mechanical model globe with an electric motor plus simple camera (right) was replaced by a digital solid-state device. Designed by the Graphic Department the new image was a fully 3D effect using solid land masses on a transparent sphere. The map data had 20,000 points and the full rotation of the computer animation takes 12 seconds.

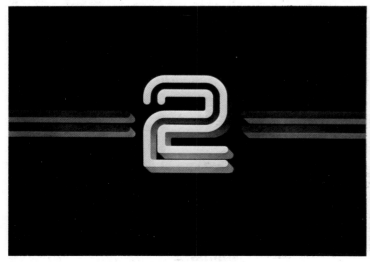

And they really were small!

Some of the earliest television receivers available to the public were only four or five inches in widths. The set on the right was described as 'the complete high-definition ultra-shortwave television receiver'. It was manufactured by Philips in 1935, and the words 'high-definition' (currently used for the anticipated transmission standard which will match 35mm film projection by broadcasting at about 1200 lines) was justified by explaining that the image would be comparable to 'a good newspaper illustration'.

However, 'A vision of the future' (drawn by Arthur Ferrier, whose latter creation 'Jane' became a television star herself see page 121) was a reality for millions of people within twenty years, in spite, or because, of the Second World War.

The magazine spread (below) from 1931, records the first live outside broadcast transmission — the Derby of that year. Did a graphic designer make a contribution then? Ironically these illustrations and graphic diagrams provide a succinct record of this historic event over fifty years later.

A VISION OF THE NEAR FUTURE.

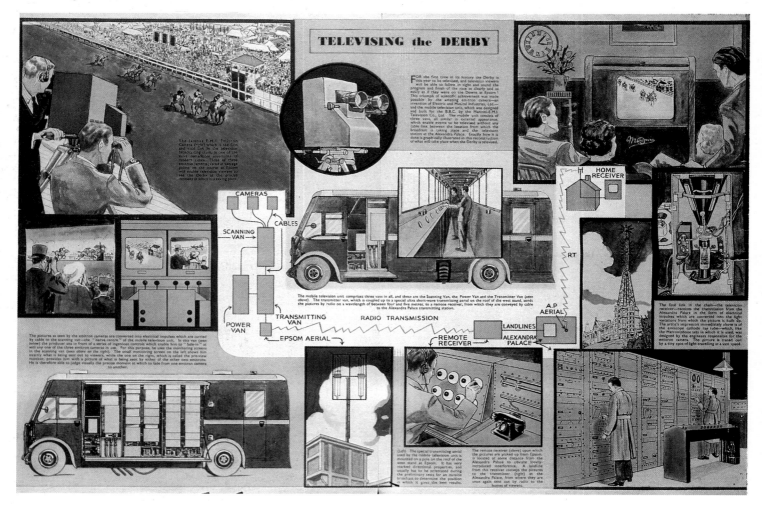

'Armchair Mystery Theatre'/ABC/UK
Repetition of words like 'Theatre', 'Armchair' and 'Magazine' occured even in the fifties when only one channel was producing 4000 programmes a year.

'Op' and 'Pop' art were strong common themes in all design in the later part of this decade and this title exemplifies the extreme simplicity of technique and 'heaviness' of early television animation.
Graphic Designer/Tony Guy

'Wednesday Magazine'/BBC/UK
Feature programmes allow the designer a freer range because, unlike drama programmes, the aim is to present a quickly recognisable image – all embrasing – and primarily to set a mood. In this 1958 title artwork and live-action of the beating of the bird's wings flowed to the staccato of a clarinet.
Graphic Designer/Bernard Lodge

'Kingsley Amis Goes Pop'/ Associated Rediffusion/UK
A surprisingly large number of graphic designers, who were part of the expansion and excitement of the mid-fifties, have remained in television. The designer of this early title went on to London Weekend Television at its inception.
Graphic Designer/John Tribe

'Top of the Form'/BBC/UK
A lively march tune fitted this simple 'jump cut' animation where the pupils spelt out the programme title with their desk tops. (1960), using still photography shot on a rostrum film camera.
Graphic Designer/Roy Laughton

'The Avengers'/ABC/UK
The boldness and extreme contrast of this 1963 design suited the very small screens and poor quality transmission of the 405 line system and the style is typical of the period.
Graphic Designer/Jerome Gask

'Darkness at Noon'/Associated Rediffusion/UK
Homage to Eisenstein's sequence on the Odessa steps from 'The Battleship Potemkin'. Here it reveals the transition to more subtle photographic images as transmission improved to 625 lines in 1964.
Graphic Designer/Arnold Schwartzman.

'City '68'/Granada/UK
Ever increasing noise of traffic and other city sounds built-up to a crescendo in a very effective way in this simple animation.
'Scratch-back' – that is scraping or painting out the art work and then running the film in reverse was the method.
Graphic Designer/Un-recorded

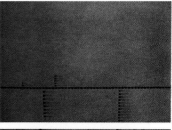

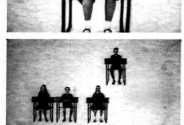

'World at War'/Thames Television/UK
Colour had arrived at the very end of the 1960s and the titles for 'The World at War' looked all the more powerful as the flames consumed the images of fear and death. (1972).
Graphic Designers/Ian Kestle and John Stamp

'I, Claudius'/BBC/UK
The pernicious intrigue of the plot was captured by the contortions of the live snake on the Roman mosaic using live-action and exploiting colour and higher standards of reception. (1976).
Graphic Designer/Dick Bailey

'The Old Grey Whistle Test'/BBC/UK
Co-ordination of the music and the vigour of the animation in this memorable sequence were very influential on graphic designers. A sense of release and exuberance were achieved by the mysterious figure. This was one of the occasions where the studio set was based on

the graphic solution. (1971).
An account of the design and production appears on page 25
Graphic Designer/Roger Ferrin

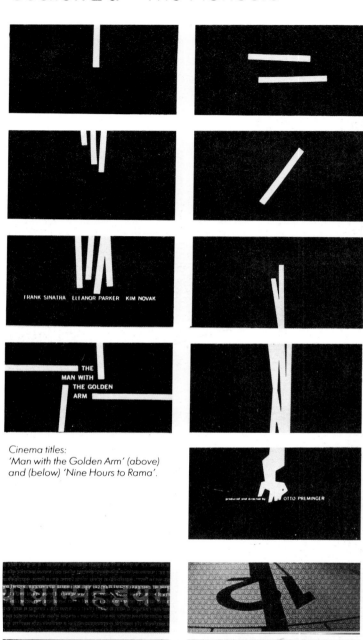

Cinema titles:
'Man with the Golden Arm' (above)
and (below) 'Nine Hours to Rama'.

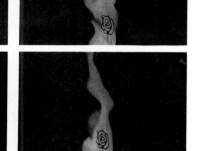

Saul Bass and film titles

Saul Bass has been regarded as the inventor of the modern film title and a strong influence on television graphic designers who grew from a small nucleus from the mid-50s. In a film on his work he said: 'I began as a graphic designer and as part of my work I created many film symbols for Ad campaigns. During that period I happened to be working on the symbols for *Carmen Jones* and *Man with the Golden Arm* for Otto

Preminger. At one point in our work Otto and I looked at each other and said "Why not make it move?" It was really as simple as that. Now additionally, I had felt for some time that the audiences' involvement with the film should really begin with the very first frame. You have to remember that until then titles had tended to be lists of dull credits, mostly ignored, or used for popcorn time. So there seemed to be a real opportunity to use titles in a new way. To actually create a climate for the story that was about to unfold.'

These animations of the early fifties were a catalyst for the graphic designers of that period. Thirty years later Saul Bass is still creating graphic work. His computer-based animations *(left and below)* based on design work for AT&T were carried out in 1985.

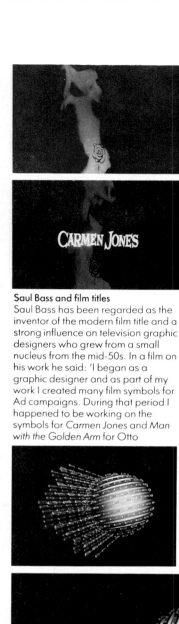

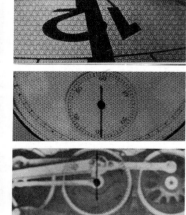

'Famous Gossips'/BBC/UK
On page 30 the graphic designer describes the background to making this opening title in the first few months of arriving at the BBC in 1965. He triumphed over his admitted inexperience in film processing techniques and 'Famous Gossips' won the first D&AD (Design and Art Directors) Silver Award for Television Graphics
Graphic Designer/Alan Jeapes

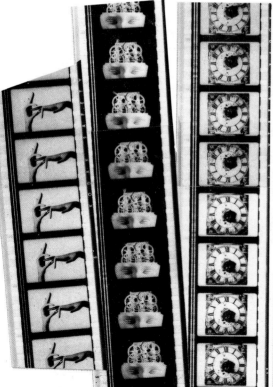

'Armchair Thriller'/ABC Television/UK
Increasingly fast, and loud, ticking of a clock for the first part – then a stark silence where the scream was expected. The title lettering appeared on the striking of a mournful knell . . . The potential of sound and images working together was realised in an effective early black-and-white 35mm film title. 1966.
Graphic Designer/Jerome Gask

The competition between BBC and ITV graphic designers was a genuine spur to both sides. This, plus the example of big budget animation applied to the cinema titles as shown opposite, encouraged high standards in much of the early television graphics.

Striving for quality and good creative standards was very difficult. There were very low budgets, and designers were limited to monochrome transmission. They were working in small sections of television production whose potential was unrecognised and poorly organised at that stage.

'Dr Who'/BBC/UK 1963 and 1973

When a video camera is pointed at monitor there is a visual response called 'feedback', or 'howl'. Bernard Lodge was one of the first graphic designers to seek out this and other new and unconventional ways of producing animation and effects on-screen. He had joined the BBC as graphic designer directly from the Royal College of Art.

The opportunity to design an opening sequence for a new childrens' series arose in 1963. On first reading he was not particularly attracted to the scripts but with the combination of television technology and conveying the storyline of space fiction and 'time warps' he encapsulated images which have endured as long as 'Dr Who' himself and pioneered design via the medium of video.

The music was synthesised by the BBC Stereophonic workshop. Later versions made for colour transmission, and because of cast changes, have maintained the mystery and technical innovation of the original. The version above was made exactly ten years after the first title. and this too exploited techniques which were new in 1973 by using computer-controlled rostrum filming.

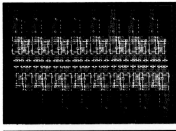

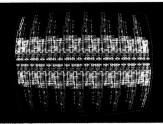

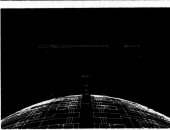

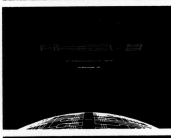
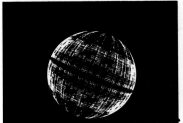

Autumn ITV/1981

An early effort to apply computer-aided animation to television graphic design. The 'wireframe' in-betweens, presented as green monitor output, were very fashionable. They were generated and drawn on a high-speed lineplotter at Middlesex Polytechnic. The detailed drawings could not have been prepared by hand. After transfer to line negatives they were filmed on a rostrum camera.
Graphic Designer/Peter Lock

'Snoot and Muttly'/1984

Animal and human figure animation has proved to be almost insuperable. One pioneering advance was made in 1984 by the Graphic Research Group of Ohio State University aided by computer time from Cranston/Csuri Productions. In 3½ minutes two very flexible bird-like creatures stalked through 3D backgrounds.
Animation/Susan van Baerle and Douglas Kingsbury

Gold Chevron/Yorkshire Television/UK

The pace and pioneering spirit of computer animation companies in America, Japan, Europe and the UK has been fascinating and of immense importance to television graphics. One of the many advanced UK companies, Electric Image, was only founded in 1984. Their partnership with Abel Associates of Hollywood has led to the development of the Digital Optical Raster Imaging System and produced moving images with the super-realism of the example below.
Graphic Designer/Jeff Parr

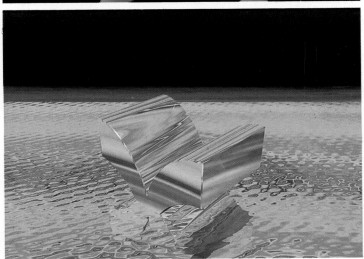

A storyboard is a plan of action. There is no one way to present the design concept for an animated sequence. The ideas, the time available, the methods of production, and even the relationship of the designer and the director involved, will be factors in deciding what method to adopt.

The most usual route is to present a storyboard. These vary enormously – from a mere scribble to many highly-finished drawings attempting to imitate almost every frame of the final animation. Some of the best ideas can be described in a few words with a wave of a hand – particularly when the graphic designer and the director have been working together for years. What type of storyboard is best? The only answer is . . . 'The one that makes the particular idea clear and understandable to the people whose job it is to decide at an early stage in the design process whether or not to proceed'. The storyboard must also assist the animator or computer programmer.

'Morecambe and Wise'/Thames Television/UK

This storyboard made the idea very clear. The cut-out paper figures were animated, as shown in the seventh frame, to imitate the famous 'dance' of the two stars. The final filmed sequence using drawn animation followed the storyboard very closely.

Graphic Designer/Ian Kestle
Animator/Eddie Radage
Rostrum camera/Peter Goodwin and Norman Hunt

'Sunday Spectrum'/ABC Television/Australia

An example of a clearly presented storyboard. The visuals would be backed-up by written or oral information production and the way the soundtrack will work.
Graphic Designer/Bill Sykes

Episode 3
FRANCE FALLS

Written and Directed by PETER BATTY

THE SECOND WORLD WAR

THE SECOND WORLD WAR

'The World at War'/Thames Television/UK
'The Second World War' was the early working title for the Thames documentary series made in early seventies when graphic designers were getting used to colour transmission. Care had to be taken to ensure the images were satisfactory in monochrome as only a minority of viewers had colour receivers. Clips of the finally completed sequence can be seen on page 51.
Graphic Designers/Ian Kestle and John Stamp
Rostrum camera/Peter Goodwin and Norman Hunt Jnr

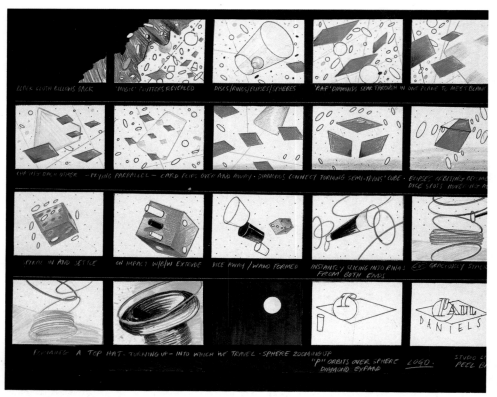

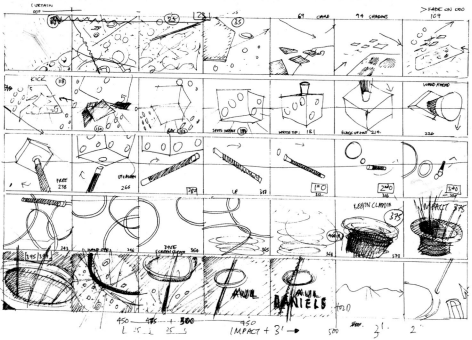

'The Paul Daniels Show'/BBC/UK

Graham Kern's storyboard for his animated title for 'Paul Daniels' underlines the point that creativity is still paramount. Computers are only as good as the ideas of the graphic designers and the people who program them.

The line version below was used to refine and explain the choreographic details of the design to the computer programer. Off-screen shots of the title reflecting the legerdemain of the star are on page 103.
Graphic Designer/Graham Kern
Computer animation/Cal Videographics

'New for '87'/ITV/UK

These four key frames were sufficient to convey a seven second animation to the ITV network Promotion Committee and give enough information to the computer animation company to estimate the cost and to decide to use ray tracing.
A frame from the final animation appears on page 23.
Graphic Designer/Mick Mannveille/Thames Television
Computer animation/Ian Bird and Paul Docherty of Electric Image

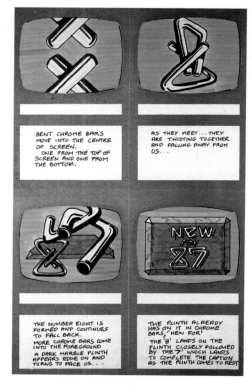

BENT CHROME BARS MOVE INTO THE CENTRE OF SCREEN. ONE FROM THE TOP OF SCREEN AND ONE FROM THE BOTTOM.

AS THEY MEET... THEY ARE TWISTING TOGETHER AND FALLING AWAY FROM US...

THE NUMBER EIGHT IS FORMED AND CONTINUES TO FALL BACK. MORE CHROME BARS COME INTO THE FOREGROUND A DARK MARBLE PLINTH APPEARS EDGE ON AND TURNS TO FACE US...

THE PLINTH ALREADY HAS ON IT IN CHROME BARS, 'NEW FOR'. THE '8' LANDS ON THE PLINTH CLOSELY FOLLOWED BY THE '7' WHICH LANDS TO COMPLETE THE CAPTION AS THE PLINTH COMES TO REST

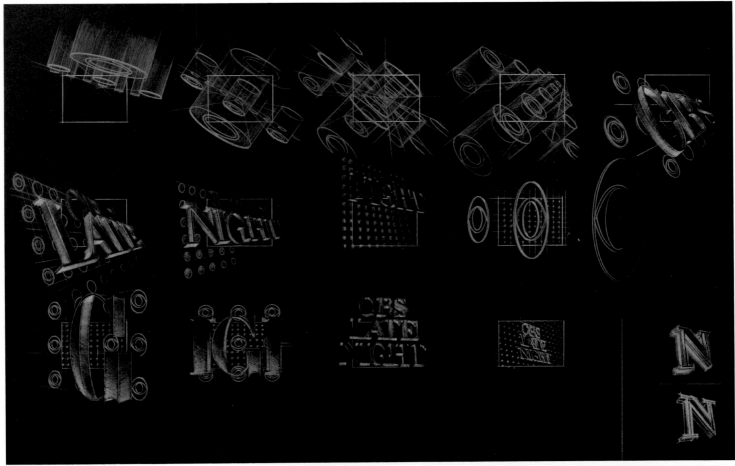

'CBS Late Night'/CBS Television/USA
This storyboard gave a highly-finished rendering of the final on-screen images. Drawing the information of each stage larger than the frame helped to indicate the flow of the movement.
Graphic Designer/Mark Hensley
for the Le Prevost Corporation

'Our World'/Granada Television/UK
The folding paper and the flight of the child's aeroplane were well animated using drawn and painted cels in this 26 second title for an educational programme for four to five year olds.
Graphic Designer/John Sharpe
Animation/Garth Wagner Animation
Music/Neil Innes

Drawing and illustration skills have always been important assets to the television graphic designer. A great many of them have chosen to illustrate their own work, although as they are often art directing a number of programmes at any one time, the amount of effort that can be devoted to this side of the production work is limited.

Freelance illustrators are commissioned for much of this output and there are many people who have specialized abilities, and knowledge, about the preparation of artwork for film animation, including drawing, painting on cels, and backgrounds.

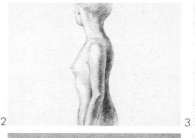

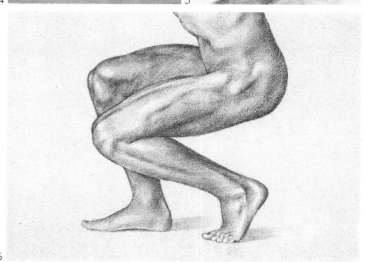

The impact and contrast of an illustration compared to a photograph is very important and very marked. (1) A watercolour painting for a 'Thames Sport' slide by Alex Forbes. (2 to 6). Drawings for an animated title by Morgan Sendall. Thames Television's 'Citizen 2000' will follow the growth and lives of a group of children who will all reach 18 in the year 2000. (7) One of many caricatures for the Thames's 'Rumpole of the Bailey' series by Rob Page. The style was based on the Ape and Spy lithographs published in the magazine 'Vanity Fair' in the late 1890s. (8) Illustration by John Leech of Granada Television for the titles of the series 'Life and Hard Times' by Charles Dickens. (9) The drama series was called 'Something in Disguise' and the graphic designer, Rob Page, used a number of drawings like this depicting the major characters. Moving cloud formation created the faces of the major characters. The actor in this example frame is Richard Vernon. The artist was Bryan Ceney. (10) Where the camera can not go . . . Drawings based on observations in court proceedings are often produced by the graphic departments for news programmes. These were made for the Australian Broadcasting Corportion by Verdon Morcom. (11) Peter Lock, a Graphic Designer at Thames Television, commissioned a range of illustrations for a Christmas on-screen promotion from Mike Brownlow.

7

Patrick
Allen
as Gradgrind

8

10

9

11

Specialised illustration and artwork, often carried out at high-speed, has always been necessary for rostrum camera filming. A great deal of experience is required to assess the amount of detail and final appearance on-screen. The output of animators has been affected by the onslaught of computer-aided methods, but recently a new wave of interest has proved that a variety of approach is more important than any one technique. Hand animators are now being employed to aid computer-aided animation!

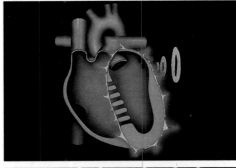

'Quincey's Quest'/Thames Television/UK
The large painted background (*above*) was only part of a London skyline prepared for a long panning shot in the opening title of a Thames's production staring Tommy Steele.
Graphic Designer/Barry O'Riordan
Illustrator/Michael Hirsch

'Soldiers'/BBC/UK (left)
To cover the range of the programmes subtitled – 'A history of men in battle', the graphic designer decided to incorporate illustrations in the opening title. The programme was transmitted at peak viewing time and was aimed at a wide audience – rather than a narrow documentary – so a very free style was requested. The illustrator used rotascope for the fast moving pieces of action, bright colours, and impressions rather than too much detail.
Graphic Designer/Liz Friedman
Illustrator/Dennis Sutton
Animation/Jerry Hibbert

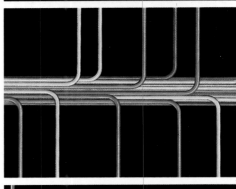

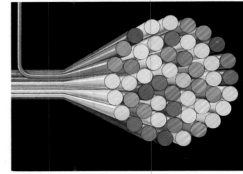

'The Body in Question'/BBC/UK *(left)*
Drawing for animation is very much in evidence in scientific series like this. The work in this case was shot on a rostrum camera but the same ability to visualise and illustrate are still essential when computers are used to present information graphics.

(Top) To show the ancient Greeks' theory that the heart worked like a furnace. (Lower illustrations) Animation to explain the nervous system.
Graphic Designers/Barbara Flinder and Charles McGhie

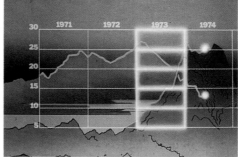

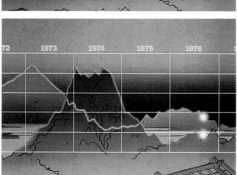

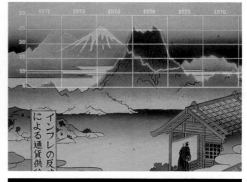

'Free to Choose'/Video Arts/UK *(right)*
Hand-painted cels and back-lit negatives were required for this insert for a documentary programme. The filmed animation explained the relationship between the growth of money supply and inflation, using the Japanese landscape in a very apt way.
Graphic Designer/Martin Lambie-Nairn
Client/Video Arts, London/WQLN Erie PA.

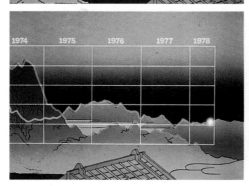

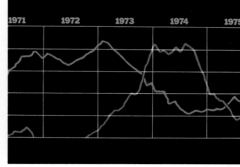

Although the most vital and commanding method of presenting the company image within television is the animated sequence, usually accompanied by music, the animated design must finally be resolved as a single still image for use on-screen and in all other forms of publicity.

Designs from stations throughout the world are shown here. Some have evolved very slowly and subtly over two, or even three decades: in other cases radical design change has occurred when the style has been altered to suit fashion, or improved technology.

(1) TV Globo/Brazil
The form of this symbol came from the early application of computer-aided animation in 1975.
Graphic Designer/Hans Donner

(2) Promotion animation/Yorkshire Television/UK
Exploitation of an ident, which proves adaptable even after years of use, is revealed in this light and imaginative summer sequence where the symbol becomes a hang-glider.
Graphic Designer/Jeff Parr
Computer animation/Electric Image

Opposite page
(3) Tyne Tees, one of the smaller stations in the UK commercial network, has retained the same symbol since it commenced transmission in 1955. **(4 and 5)** London Weekend Television have also kept very closely to their original design from the time they first operated the franchise in 1968. The recent 'new look' in 1986 made only very minor changes. Television New Zealand **(6)** uses this symbol on-screen on stationery and in publicity material. Graphic Designer/Roy Good. The WDR symbol **(7)** comes from the late 1960s when many radio and television stations used radiating lines and concentric circles to suggest broadcasting. The Japanese station FTC **(8)** a bold mark, again from the 60s — ideal for the less refined transmission standards of that period. **(9)** This is a computer-generated version of the Australian Broadcasting Corporation's widely applied ident. A strong graphic image **(10)** linking Switzerland to television production in 1970. The final frame **(11)** of a recent computer animation, designed and produced by Robinson Lambie-Nairn Limited, presents the company 'trademark' for Scottish Television. The versatility of the unchanging Yorkshire symbol **(12)** as demonstrated on the opposite page. The search for renewal, and adaptation to fashion and technical changes, for BBC2 **(13 and 14)** — from 1964 and 1986.

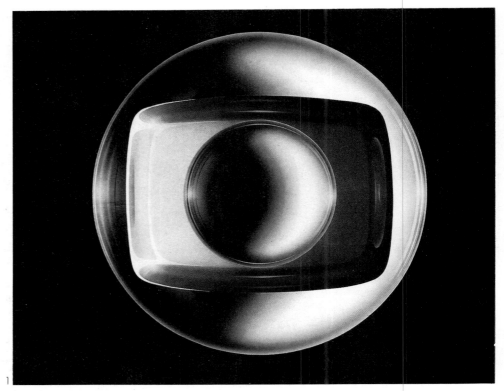

1

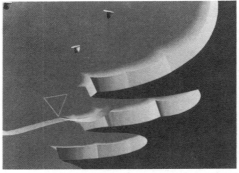

2

3

4

5

6

7

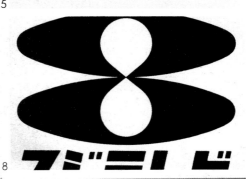

8

9

10

11

12

13

14

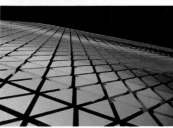

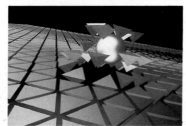

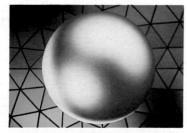

The expansion of television stations and networks throughout the world has produced a vast range and variety of animated ident sequences. By their power and constant repetition they have become instantly recognisable to audiences of many millions. Very few idents, in any part of the world, are now produced without using computer-aided animation. One of the very first companies to do so was TV Globo, transmitting to 70 million Brazilians in the mid-1970s and this ident had the virtue of being presented in many forms.

TV Globo/Brazil/1986
This is a recent development of their computer animated indent of the 70s.
Art Director/Hans Donner
Graphic Designers/Hans Donner and Nilton Nunes
Computer programer/Lucia Modesto

Graphic designers have been presented with new markets by the changes in the broadcasting systems. Satellite and cable networks have looked to graphic designers to give them visual 'personality'. The demand for good design to launch into these areas has been an important new stimulus.

Channel 4/UK
The aim was to represent the diversity of Channel 4's programme sources. Although only produced in 1983 the computer work had to be carried out in the USA.
Graphic Designer/Martin Lambie-Nairn.   Production/Robinson Lambie-Nairn Limited   Computer animation/Bo Gehring Aviation, Los Angeles

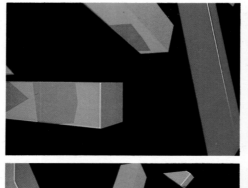

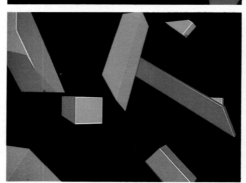

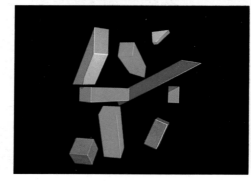

NOS/Holland
This 12 second sequence introduces all NOS programmes and was one of the first station animations to be produced on the Ampex Cubicomp.
Graphic Designer/Ron van Roon
Computer animation/The Frame, Amsterdam and NOS

Granada Television/UK
Although this animation is not the adopted station ident the sequence is typical of those that reinforce the company style. This one up-dated the old Granada symbol in computer generated fashion in 1985.
Graphic Designer/Peter Terry
Computer animation/Peter Florence/ Digital Pictures

Kron-TV/USA
San Francisco's greatest landmark was animated using PDI's in-house animation software.
Graphic Designers/Bruce Lindgren and Judy Rosenfeld
Computer animation/Thaddeus Beier of Pacific Data Images

Mirrorvision/UK
In this on-screen indent for a cable television company the mirror moved on its axis and reflected the colours and textures around it.
Graphic Designer/Bob English of English Markell and Pockett
Computer animation/Paul Docherty of Electric Image

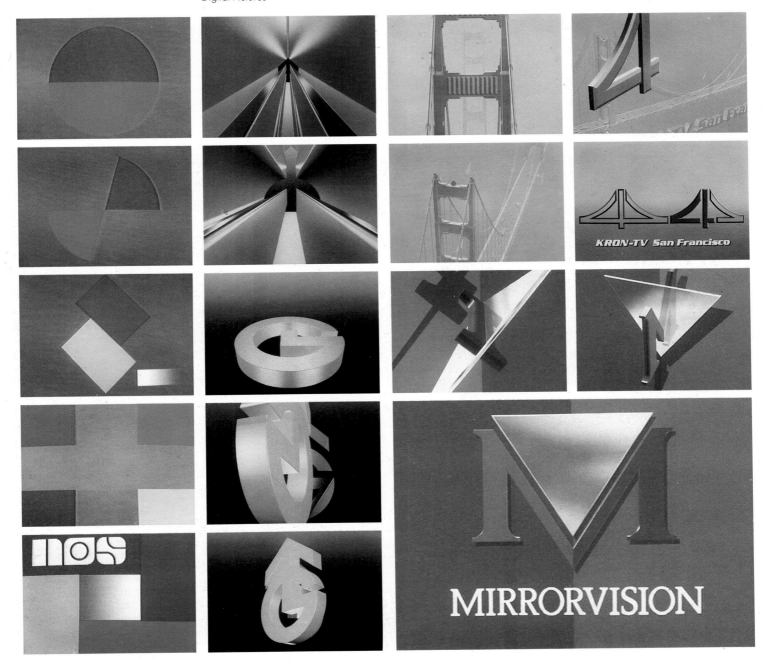

(Top right) Simplicity for monochrome transmission in the days of **Associated Rediffusion**. Graphic Designer/John Tribe and a more recent colour slide from BBC Scotland. (Above) Early slides for **ABC Television** for 'The Avengers'. Graphic Designer/John Stamp. (Below) More black-and-white promotion slides from the **BBC** during the 1960s. Note none of these proclaim the station's name.

With more networks the identity of the station becomes important. (Above) Two promotion slides from 1969 for **Granada** by John Leech and (below) promotion slides for BBC2 and the open university.

Information about forthcoming programmes is still given in the form of single illustrated captions, or lists of programme titles – known as 'menus' or 'rundowns' – as part of television transmission. They are required in large quantities and they are designed and produced as quickly as possible. Because they are on screen for only a few seconds they must get an impression of the programme across very quickly.

Most of these shown here were drawn, painted, or based on photographic artwork. Those from Channel 4 are carried out entirely on a digital paint system and transmitted via an electronic stills store.

These slides from **Thames Television** demonstrate the use of conventional artwork and photographic origination as well as various ways of displaying the company's ident.

Channel 4 in the UK has come upon the scene more recently and is now one of the first companies to produce all its promotion slides using an electronic digital paint system. All these were designed and produced by Simon Broom.

'Menus', or 'Rundowns', from **London Weekend Television** present lists of forthcoming programmes. Their design and production is an everyday task for graphic designers. Electronic caption generators have streamlined the composition and display of this important but routine work. For many years these were handwritten or typeset.

Two promotion slides for drama programmes for Harlech Television – the ITV company which has the responsibility for Welsh language transmissions.
Graphic Designer/David James

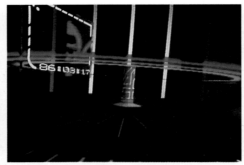

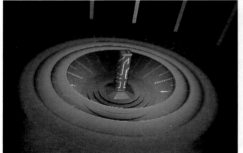

Channel 4 does not make programmes but it has set-up a small graphic design unit and it is their job to design and produce all the company's on-screen promotion work. They have made excellent use of the numeral '4' and present it in many original ways.

New Year 1985/Channel 4/UK Videographics
The frames above are from the introductory sequence to the New Year trails and those below are from the exit. The animation was generated on a VAX 11/750 and incorporated texture mapping used Paintbox images.

Graphic Designer/Marc Ortmans
Computer production/Gareth Edwards of Cal
Videographics/London
Music/Nick Bicat

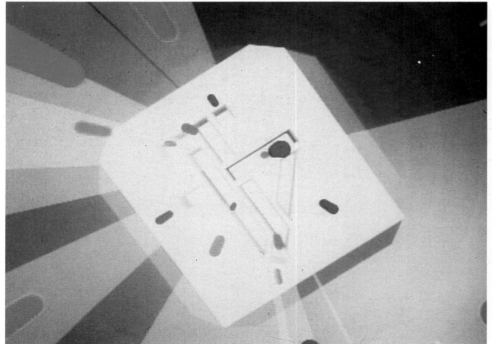

Autumn 1985 ITV Network
The task of designing the network on-screen promotion material in the Independent Television is shared by the five largest contractors in turn. The length of promotion animations has been reduced over the years. Originally this theme, designed by the Graphic Design Department of Yorkshire Television, was more complex but with only seven seconds the storyboard was simplified at the computer facility house.
Graphic Designer/Graham Smith
Computer animation/Paul Docherty of Electric Image

 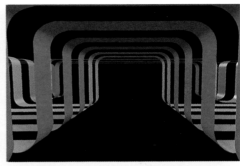

'New Season 1986'/Thames Television/UK
There were only four weeks between brief to the first transmission of this computer animation used to 'package' Thames autumn programmes. The very strong movement lasted only five seconds and was repeated many times throughout the day.
Graphic Designer/Rob Page
Computer animation/Tony Ford of Virgin Computer Graphics

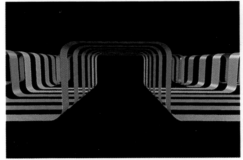 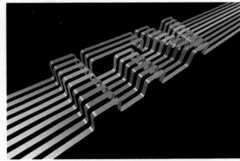

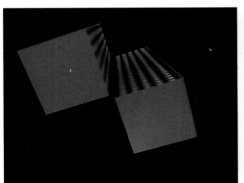 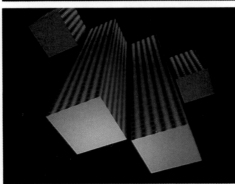 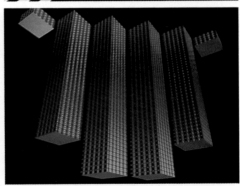

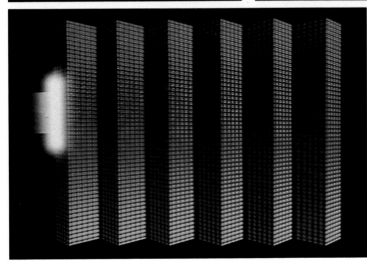 

CBS Movie Presentation/USA
This animation was one in a series of four promotions for film presentations. Parts of its 25 second duration were designed to be pulled out as 8, 6 and 3 second 'bumpers' to be used before, or after, commercial breaks.
Graphic Designer/Mark Hensley for The Leprovost Corporation
Animator/Olen Entis
Production company/Pacific Data Images

 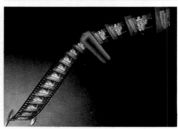 

'The Entertainers'/London Weekend Television/UK
The hand-drawn animation team which made 'The South Bank Show' titles so vital were responsible for this sequence showing many of the stars at LWT.
Graphic Designer/Pat Gavin
Animation/Pat Gavin and Jerry Hibbert

  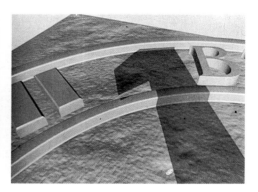

'Autumn 1984'/BBC2/UK
The only instruction to the designers
for this animation was 'Create
something moving forward', and the
music was provided. There was a
strong resistance to using autumn
leaves or any seasonal references.
Graphic Designers/Liz Jones and Fen
Symonds
Computer animation/Digital Pictures

'Days of the Week'/BBC/UK
The gnomon on the sundial was
formed by the '1' of BBC1. Specially
programed texture and 'lighting' – to
simulate uneven lead – gained the
compliment that the animation was
filmed from a model and did not have
the immediate 'computer look'.
Graphic Designer/Harry Donnington
Computer animation/Digital Pictures

Christmas 1985/BBC2/UK
Traditional cel animation and
traditional seasonal images came
together in this on-screen promotion.
Graphic Designer/Fen Symonds
Animation/Brian Larkin/Animation
People

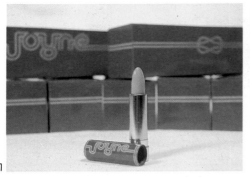

1

2

Away from the titles and award-winning side of graphic design there is the day-to-day production of signs, printed matter, newspapers, books, packaging and posters. All these have to be designed, printed, and produced for dramas, sit-coms, variety shows and even current affairs programmes. The range in period and quantity is limitless and the examples here try to echo the almost bizarre nature of some of the requests — 'A toe tag for a corpse from the early 1930's'!

Dressing a complete series set in a hospital, a police station, or in a Victorian seaside town, can give some idea of the quantity and variety of signwriting which is required in television production. Careful observation and research to achieve authenticity is vital.

A range of cosmetics (1) was featured in a BBC programme and no existing brand name could be used. Some of the typical graphic 'props' in (2) might be burnt or destroyed in some way during the action of the drama. Newspapers from all periods and countries are a constant product. Antique props were needed in large quantities (3) for the HTV Wales production of 'Return to Treasure Island'. Graphic Designer/Ray Lambert. A children's programme made a joke (4) and graphics provided this visual. Graphic Designer/Alex Forbes. (5) Could this be a rival to 'Private Eye'?/A magazine for 'Lytton's Diary' by Thames Television. Graphic Designer/Rob Page. In 'The McGuffin' material was needed to flesh-out the background of a writer on the cinema. Jackets for the books (6) and (7) he had written, posters and other film ephemera were an essential part of the play's references. BBC/Graphic Designer/Mina Martinez. A reconstruction (8) of an otherwise unobtainable music hall poster.

3

4

5

6

7

8

Signs and signwriting
Making signs and signwriting for locations, and for studio settings, is a constant but very varied element of a graphic design department's daily work.

'Weekend World'/London Weekend Television/UK *(above)*
The truly international breadth of LWT's current affair coverage was emphasised by a much more pictorial use of computer-generated images than those previously used.
Graphic Designer/Martin Lambie-Nairn
Robinson Lambie-Nairn Limited
Computer animation/Digital Pictures Limited

'World in Action'/Granada/UK *(opposite)*
When this title was redesigned in 1980 the original static image of the mathematically proportioned man by Leonardo da Vinci was retained but enlivened with a flying white spot. News events were excluded, only the simplest suggestions of action were visible.
Graphic Designer/Ray Freeman
Rostrum camera/National Screen Services

'What the Papers Say'/Granada/UK *(right)*
One of the longest running programmes on television had an opening title using views of Fleet Street buildings. Now that so many printing bases have moved away the title has changed. The design and production of this sequence is described on page 34. (1986)
Graphic Designer/Peter Terry
Film animation/Sprockets, Manchester

Arts and feature programmes rely on graphic design a great deal to portray mood and theme, and unlike drama titles, there is much greater scope for a personal and creative approach. Some of the most energetic and original solutions have derived from this area.

'The South Bank Show'/LWT/UK

'Make the Arts look accessible' was Melvyn Bragg's brief when this long-running series was first discussed. From that request Pat Gavin of London Weekend Television produced an animation with an infectious vitality that has seen a number of revisions to the basic theme. The reference to the creation from the Sistine Chapel plus a creative spark was inspired visual thinking – rarely equalled. The sixteen letters of the title coincided with perfect serendipity with the final sixteen beats of the music – a setting of 'Variations on a theme by Pagannini' by Andrew Lloyd Webber.

The graphic designer's account of the design and production of these titles appears on page 26.

Graphic Designer/Pat Gavin
Animation/Pat Gavin & Jerry Hibbert

'Ghosts In The Machine'/Channel 4/UK

Title sequence for a series on experimental video art.

The sequence used as much current video technology as possible given budget and time constraints. The following effects/ equipment were used – Dubner, Paintbox, Mirage, ADO and all live-action material was specially shot using a video camera.

Design and Direction/Richard Markell/ English Markell Pockett Production Company/Illuminations Editor/Rob Banachie

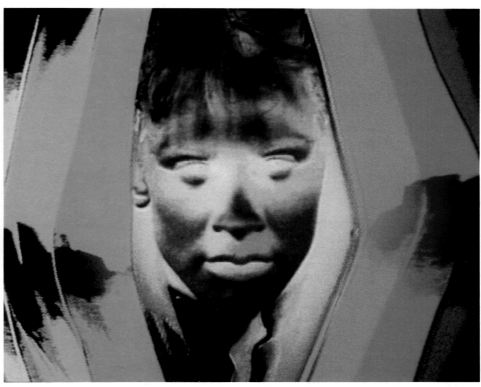

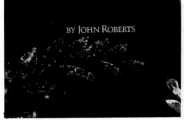

'The Triumph of the West'/BBC/UK
The model of the victor's laurel was articulated so that the leaves appeared to decay and fall away. The leaves were real bay leaves with gold applied by a vacuum technique. The working unit was shot against a black velvet background. (1984)
Graphic Designer/Alan Jeapes
Modelmaker/Steve Wilsher, Creating Effects.

'The Italians' ABC/Australia (below)
This filmed title was inspired by the photomontage themes found in the Futurist movement and produced a 'pageant' for the series made by the Australian Broadcasting Corporation. Italian immigration to Australia and the subsequent history of those people was examined in these programmes. The graphic designer's description of this sequence appears on page 35.
Graphic Designer/David Webster

'Crime Inc'/Thames Television/UK (opposite)
A short dramatic burst of face-to-face interviews, known as a 'hooker', was run at the beginning of each programme in this series before this explosive title. The music stresses were emphatic and their timing very precise. The designer's account of this title appears on page 28.
Graphic Designer/Lester Halhed
High-speed film animation/Douglas Adamson of Stewart Hardy

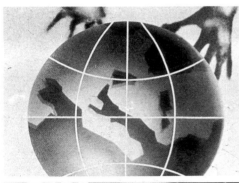

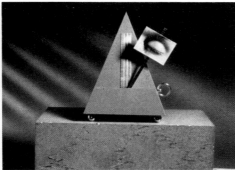

'Forty Minutes'/BBC/UK
The programme editor, Eddie Mirzeoff, asked the graphic designer to avoid the obvious elements – clock faces, clock mechanisms and hourglasses. He also asked for the sequence to be continuous. The designer's proposal was to borrow Man Ray's surreal moving eye from 'Destroyed object' 1932. The eye was made to open as the movement occurred.
Graphic Designer/Graham McCallum
Model animation/Gillie Potter
Music/Carl Davis

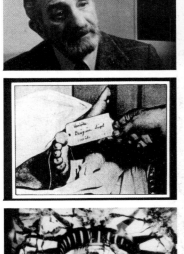

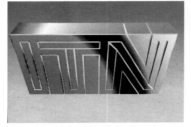

Channel 4 News/ITN/UK (above)
This news programme produced by ITN first went on air in 1982. Three years later a revision was made using the fashionable reflective metal surfacing provided by computer developments.
Graphic Designer/Lesley Friend
Computer animation/Gordon Plant of Cal Video, London

'Jornalda Globo'/TV Globo/Brazil (below)
The station identity, launched in 1975, has extended to most programme titles. The company attracts an audience of over 70 million.
Graphic Designers/Hans Donner and Nilton Nunes
Computer animation/Rogéruo Ponce

'ABC News'/ABC/Australia (below)
Models of the earth and radar dish were made in ABC's Scenic Department. The animation is 30 seconds long – excluding the news headlines.
Graphic Designer/Terry Dyer
Model motion-control/Zap Productions

'24 Hours'/CBC/Canada (below)
Recently created for the Winnipeg station this title is another example of Canadian graphic design where the animation has been achieved 'in-house' using a CG1 Image II computer.
Graphic Designer/Gordon Morris

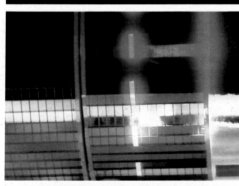

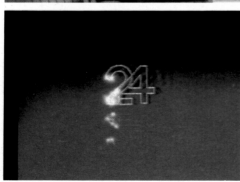

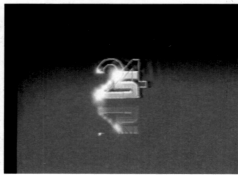

'Eye Witness'/NZ Television/New Zealand (right)
Digital effects and computer-animated lettering present the Auckland skyline in this news title.
Graphic Designer/Darryl White

'Headline News'/Atlanta/USA
The components of this nine second title for 'Who's News?' were prepared on the Aurora 125 digital paint system and combined in video editing using Ampex ADO and a Grass Valley 300 switcher. Headline News is a 24-hour 'news only' station.
Graphic Designer/Missy Fogel
Art Director/Frances Heaney

'WCBS-TV'/New York/USA
The Art Director of this CBS station, Scott Miller, praises the introduction of digital paint systems.
'Only designers and artists can fully utilize the power of these expensive machines. With these tools designers are regaining control over visual effects and the means of graphic production. At

most tv stations the centre of graphic identity is the local news.'
Recent major revisions to the WCBS News graphics style are reflected in the examples below and the background to graphic design at CBS is on page 31.
Graphic Designers
Top three/Robb Wyatt
Lower two/Milo Hess

Weather information is usually presented at the end of news programmes, and produced by the graphic designers on the news teams. Styles from four UK Stations are shown. Straight-forward information graphics from Paintbox at Thames. Two light-hearted approaches for early mornings from TVam. Computer-controlled meteorological information via the BBC, and evocative backgrounds, produced for character generator overlays, from London Weekend Television.

'Thames News'/UK

'Thames News' transmits to 9 million Londoners and the five bulletins throughout the day are served by three graphic designers.

Since early 1986 the programme has had its own Video Graphic Workshop with a 'Paintbox'. The current graphic style is reflected in the studio *(right)*. *(Below)* Three conventionally produced captions *(on the left)* and three electronic examples. Without resorting to photographic processes the volume of work and the speed of production has been vastly improved.

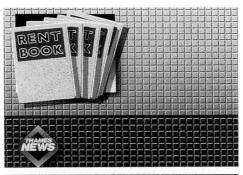

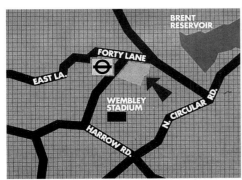

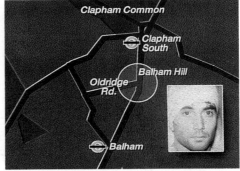

'Elections were in many ways the beginning of information graphics on television'.
Chris Long/Manager of Computing and Graphics at Independent Television News/UK.

The spur for ITN to purchase their first computer was the General Election in 1974 and the results were displayed directly on-screen using a VT80 computer **(1)** for the first time in British television. Both ITN and BBC spent a lot of time and money in improving graphics via their news departments for the 1980 American Presidential elections. This upgrading remains as part of the daily newscasts.
A reminder of the old form of studio presentation is shown in the pictures below **(2 & 3)** of a London Borough Election some years ago. Cut-out cardboard shapes were applied by hand in front of a studio camera!

'General Election' and '1980 American Election'/BBC/UK
This refined typographic style was the result of months of planning and the exemplary detail has achieved much of the finesse of the printed page. The off-screen shots of computer output **(4 & 5)** show the UK Election of 1981 and **(6 & 7)** show the BBC's presentation of the 1980 American Election.
Graphic Designer/George Daulby

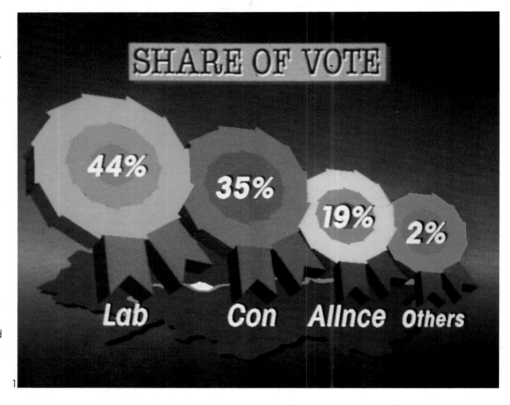

1

2

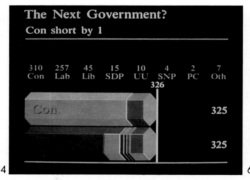

4

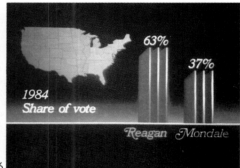

6

3

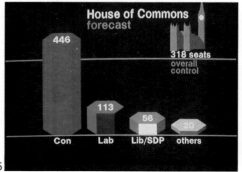

5

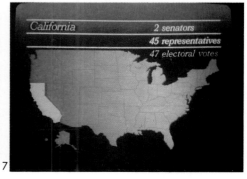

7

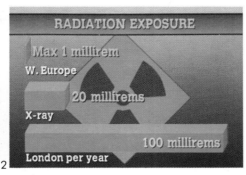

ITN/Switch to electronics
At ITN a team of graphic designers now work alongside computer programers to produce animations, as those above (1 & 2), within a few hours.

The ITN news graphics computer equipment now consists of two VT80s – for use on every day rather than specials – four Paintboxes, an Aston IV character generator and a Harris frame store.

Manchester air crash/ITN/UK
The value of information graphics was shown by the ITN coverage of the official report of the Manchester air crash disaster. There was no filmed record of the scene and the graphic designers pieced together as much information as possible (3, 4 & 5) and made last minute amendments when the report was released. Their work received commendation from the Royal Television Society in a topicality award.

Large quantities of still graphic images are prepared ahead of the news bulletins, 'stacked' in electronic stills stores and then recalled to be keyed-in behind the news presenters (6, 7 & 8) using single colour 'Chromakey' backgrounds – usually blue – as masks.

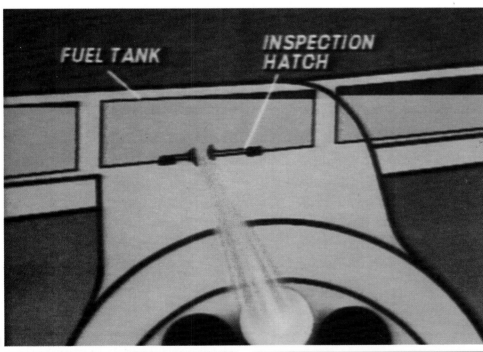

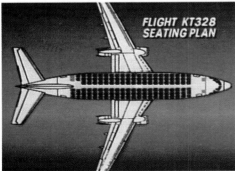

'ABC News'/New York
A collection of Paintbox still frames from the ABC programme 'World News Tonight' reveals a very distinctive American style. Objects and type, often heavily shaded, tend to be large within the frame. Compare these with the early AVA designs on page 128. Director of Graphics/Ben Blank

'Channel 4 News'/UK *(6 and 7)*
At the start of 'Channel 4 News' four years ago, ITN decided to use the VT80 computer for animations every night. This was so successful the system quickly developed to provide similar animations for all ITN bulletins. The histograms show nightly financial updating, and a novelty piece on electricity production power sources.
(8 and 9)

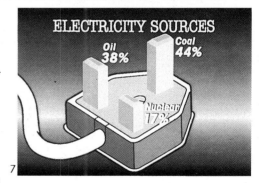

The ITN graphic service has a complete world map on a data base. Any outline can be selected and automatically generated in the style chosen for either 'ITN News' or 'Channel 4 News' information.

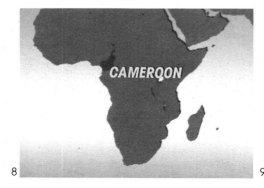

'ABC Television'/Australia

All the 'ABC News' graphic work in Sydney is now created on Quantel's Paintbox and stored on the digital library system. A freeze frame of the set (1) has been stored, with the graphic format superimposed, in the Paintbox memory. This forms the base for all the Chromakeyed graphics. The long portrait shape deliberately overlaps the presenters shoulder to maintain picture depth.
Art Director/Kathryn Day

The programme style is established in the special generic designs (**2**, **3**, and **4**) where Times Roman Bold, with pale gradations, tie-in the opening titles.

Full frame formats (**5**, **6**, and **7**) have character generated type superimposed over Paintbox backgrounds from the digital library system. Colour and layout for these 'prepacked' – as opposed to the programme – graphics is left to the individual designer within the story requirements.
Format/Michael Murray
Graphic Designers:
2 Michael Murray
3 and **6** Robin Johnson
1, **4** and **5** Kathryn Day
7 Michael Murray

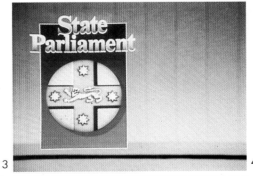

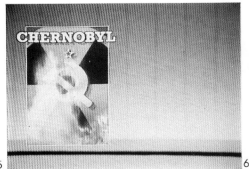

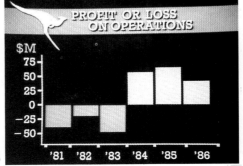

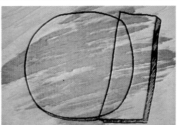

'Are You Awake Yet'/TVam/UK
A rising sun, a ship, a ball and a kite, cut between shots of small kids leaping out of bed, made the letter shapes required. Breakfast shows have needed to pay special attention to the young audience.
Graphic Designer/Geoff Hurst

'Splash!'/Thames Television/UK
Magazine programmes to attract young teenagers have to cater for fast-changing fashions. The aim of this very complex title are described by the graphic designer on page 33.
Graphic Designer/Barry O'Riordan
High-Speed camera/filming/Oxford Scientific Films
Rostrum camera/Norman Hunt and Vic Cummings
Animation/Reg Lodge

Balão Mágico'/TV Globo/Brazil
The high standards of graphic presentation at TV Globo are applied to their children's output. Here the sets and titles set the mood for the Magic Balloon. Hans Donner's account is on page **23**.
Art Direction/Hans Donner
Design/Nilton Nunes, Ruth Reis, Luc Boels, Ricardo Navenberg, and Alvaro Barada

'Roland's Rat' Race'/TVam/UK
The graphic designers at TVam transferred Roland, the puppet character, to cartoon animation for the title sequence, using their in-house Paintbox system.
Graphic Designer/Bob Herbert

Most television services include programmes made for adult and child education. Providing graphic material for these programmes is providing an ever-increasing source of very detailed work, often produced to stringent budgets. In the UK some of the graphic designers at the BBC are involved in a quite separate unit working solely for The Open University at Milton Keynes and samples of their work are shown in items 1 to 6. The Graphic Design Manager is Susan Dix.

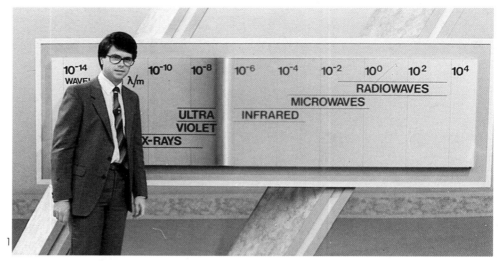

1 Studio graphics for 'The Physics of Matter' on panels, where the presenter can explain the visual information, are still a widely employed method in educational programmes as they are direct and economical.
Graphic Designer/Linda Davis

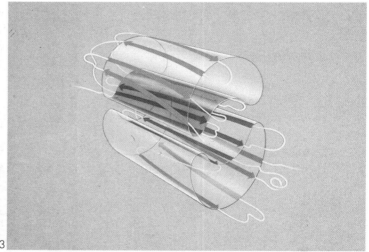

2 The educational output requires its own promotion material as this slide for 'The Rise of Modernism in Music' illustrates.
Graphic Designer/Rose James

3 Animation is extremely valuable for explaining complex ideas: this frame is from cel animation for 'Biochemistry and Cell Biology'. Graphic Designer/Gary Shorland

4 An example of illustration work for a series on geology. Graphic Designer/Paul Bond

5 Computer animation is an ideal adjunct to the mathematic and scientific content of much educational work. Here it calculated and produced an animation of the DNA helix.
Graphic Designer/Brendon Ross
Computer animation/Jilly Knight/
Electronic Arts

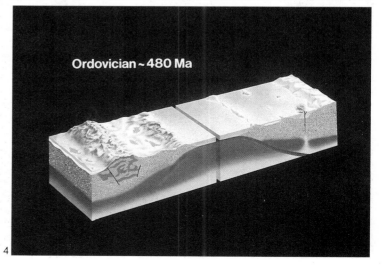

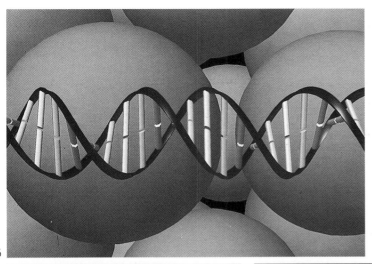

5

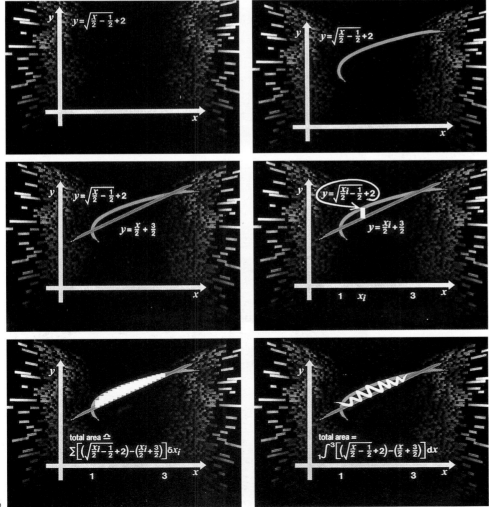

6

6 These frames from a film sequence on calculus reveal just how exacting much of the graphic work for academic studies can be.
Graphic Designer/Brendon Ross

'Scientific Eye'/Yorkshire Television/UK
Yorkshire have a Documentaries and Education Graphics Department. One of their networked productions is a series for children on science. The first two illustrations show the programme symbol and logotype and the others are from animations for programme topics. Tog values for 'Eskimo Clothes' and a programme on 'Gravity' refering to Galileo and Isaac Newton.
Graphic Designer/Programme symbol/David R Gledhill
Eskimo clothes/Illustrator/Gavin Macleod

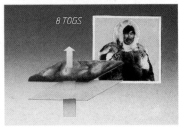

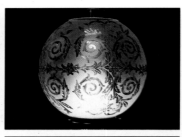

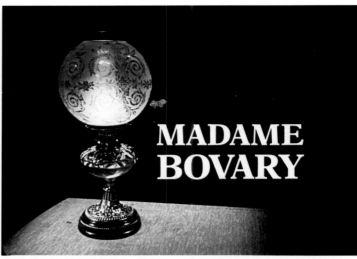

'Madame Bovary'/BBC/UK
The fate of the moth drawn to the flame of the lamp graphically reminds the viewer of the fate of Emma in Flaubert's tragic novel. Filming the title occured when there were no moths 'in season'. The erratic flight of the creature was very convincingly achieved using hand-painted cel animation. (1975).
Graphic Designer/Stephan Pstrowski

'Rock Follies '77'/Thames Television/UK
A second series of this musical drama told the story of a three-girl Rock band and their search to get their name in lights. The 1975 version, and this, were the first to rely entirely on the animation using back-lit cels. The music was written exactly to the designer's wishes by Andy Mackay of Roxy Music.
Graphic Designer/Rob Page
Rostrum Camera/Peter Goodwin and Norman Hunt Jnr

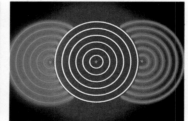

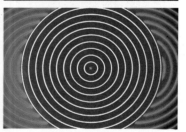

'Secret Army'/BBC/UK
Alan Jeapes's title was used on all three series of this 13 episode drama. He did not formulate his theme until he had listened many times to the music composed by Robert Farnon. When his proposals were made he was pleased that the producer, Gerard Glaister, accepted such a simple and seemingly low-key solution to a dramatic subject. Jeapes was awarded the British Academy of Film and Television Arts Award for Television Graphic Design for Drama title in 1979 and a D&AD Silver Award in the same year.
The stills were selected by the graphic designer on location in Belgium and shot on a computer-controlled rostrum to obtain the expotential zooms.
For the end credits the similar compostions were used with a final shot of the open sea – a sign of escape from occupied Europe. (1978)
Graphic Designer/Alan Jeapes

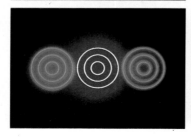

'The Wednesday Play'/BBC/UK
A black-and-white filmed titled for a drama series from the days of simpler production systems.
Graphic Designer/Oliver Elmes

'East Enders'/BBC/UK
Slow panning across aerial photographs of London's docklands and eastern areas with a strong music theme carries this most successful 'soap opera' into over 20 million homes on each transmission.
Graphic Designer/Alan Jeapes

'The Loves and Lives of a She-Devil'/BBC/UK
This recently transmitted series was a macabre fable of a plain housewife whose husband falls in love with a beautiful and rich romantic novelist. An account of the making of the title and the special graphic props is given on page 27.
Graphic Designer/Michael Graham-Smith
Live-action filming/Oxford Scientific Films
Computer animation/Electric Image

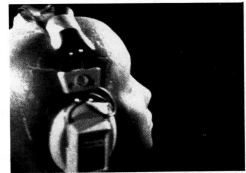

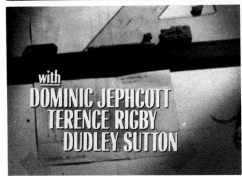

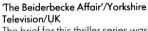

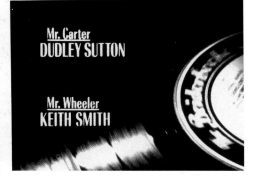

'The Beiderbecke Affair'/Yorkshire Television/UK
The brief for this thriller series was 'To embrace everything in the scripts – but give nothing of the plot away'. The graphic designer's report on the design of the titles for this series, written by Alan Plater, is on page 24. She used 35mm live-action filming in a special version of the hero's room built in the stills studio at Yorkshire Television and strove to shoot objects which had movement without becoming too strident, or obvious. The last frame show here is from the end credits.
Graphic Designer/Diana Dunn

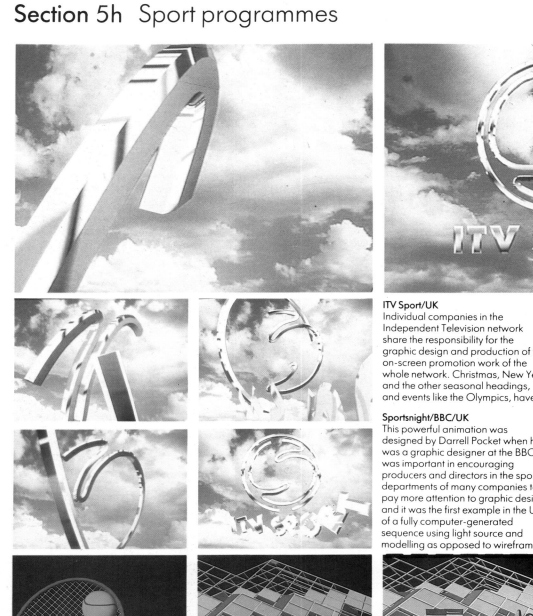

ITV Sport/UK

Individual companies in the Independent Television network share the responsibility for the graphic design and production of the on-screen promotion work of the whole network. Christmas, New Year, and the other seasonal headings, and events like the Olympics, have special animations and these are transmitted on all the stations in the network. This computer-generated sequence for the sport presentations was undertaken by London Weekend Television and first used in early 1986.
Graphic Designer/Chris Hart
Computer animation/Digital Arts

Sportsnight/BBC/UK

This powerful animation was designed by Darrell Pocket when he was a graphic designer at the BBC. It was important in encouraging producers and directors in the sports departments of many companies to pay more attention to graphic design and it was the first example in the UK of a fully computer-generated sequence using light source and modelling as opposed to wireframe.

From the first appearance of the floating video camera high above the floodlit stadium the pace, from one type of sport to another, was fast and very sharply edited. The hardware was two Vax 750 computers, a Gems frame buffer, a Dicomed D48 high-resolution film recorder and a Pyramid 90X/Iris.
Graphic Designer/Darrell Pocket
Computer animation/Electronic Arts

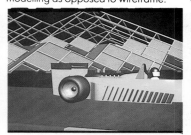

Headline News/Atlanta/USA

This opening and closing title lasts a bare seven seconds. In that time it links the six main sports the programme covers.

The graphic designers working in this all-news television station in Georgia used live-action video keyed over moving background of an airbrushed sky and drawings of each type of playing surface.

A Grass Valley 300 switcher and Ampex ADO were the main equipment employed.

Art Director/Frances Heaney
Graphic Designers/Stan Anderson and Cindy Varnes

Coppa del Mondo/RAI/Italy

The Italian network RAI commissioned the design and production of this complex 90-second animation from a London facility company and three months of VAX computer, Paintbox, and post-production work were required.

Graphic Designer/Terry Hilton
Paintbox design/Peter Kavangh
Music/Nick Glennie-Smith
Computer production/CAL Videographics

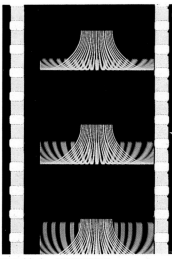
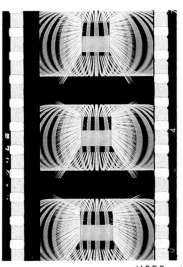
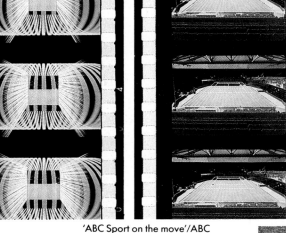

'Wimbledon 85'/BBC/UK
Film techniques are still being employed alongside computer animation. A wide-angle view of the court was combined with back-lit negatives and shot in 35mm film on a computer-controlled rostrum camera. The six second sequence for the annual Wimbledon title was used as a lead into live-action shots.
Graphic Designer/Martin Foster
Rostrum Camera/Colin Hancock

'ABC Sport on the move'/ABC Television/Australia
The 'crowd' effect was achieved by exploding colour bars with a Mirage. A small section was chosen and then texture mapped, via Paintbox into a Bosch FGS4000. The sequence is a sport promotion — not a programme title.
Graphic Designer/Julian Eddy
Computer animation/Video Paintbrush Company

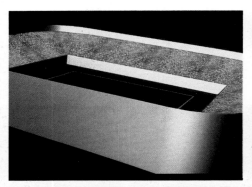
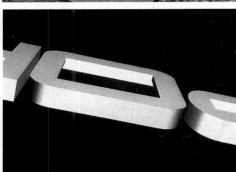

'Grandstand'/BBC/UK

The viewer appears to be driving a racing car at very high-speed between placards. At a number of points barriers are broken to reveal brief shots of sports stars in action. Finally the camera pulls out to take a high view point of the grandstand as it grows with the camera movement to display the programme's title. Produced in June 1986.
Graphic Designer/Rod Ellis
Computer animation/Chris Fynes and John Speirs of Crown Computer Graphics

'World Tenpin Bowling'/CBC/Canada

CBS are among the first in-house graphic design departments to use their own animation equipment. With cheaper computer systems needing less programming and which are 'user friendly' to the point of total intimacy, the practice will grow. They have two 'Image II' Systems operating with three shifts up to 22 hours a day. This animation took 24 man hours to complete excluding the overnight recording time.
Graphic Designer/Christian Castel

 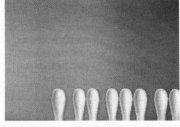  

WORLD TENPIN BOWLING

'We Love TV'/London Weekend Television/UK
This quiz programme revived the fifty-year-old ident in the 'BBC News and Newsreel' title shown on page 46. In this version, made in 1983, the pulsing rings were keyed from computer animation into Paintbox with the re-drawn mast from the old Alexandra Palace transmitter combined with Mirage effects.
Graphic Designer/Chris Hart
Computer production/CAL Videographics

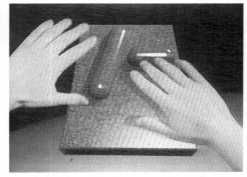

'The Krypton Factor'/Granada/UK
Made for the tenth anniversary of this series the marble slab and the mysterious elements – echoing some of the puzzles in the programme – were computer-generated but the hands were real. These elements were combined using Ultimatte.
Graphic Designer/Murry Cook
Computer animation/Mike Milne of Electric Image

'Entertainment'/CBC/Canada
A title which avoids the over-polished appearance of so much of the present computer-based work but was produced on a computer within the graphic design department in Toronto for a midday show.
Graphic Designer/Christian Castel
Art Director/Gerard Bueche

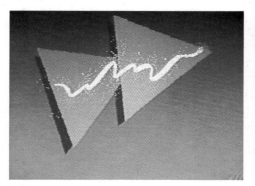

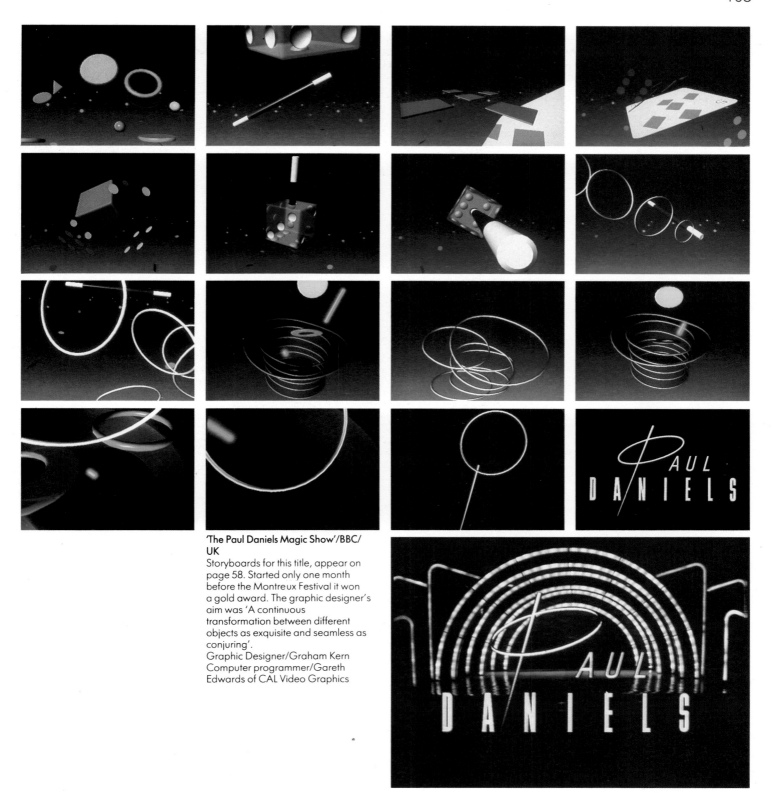

'The Paul Daniels Magic Show'/BBC/
UK
Storyboards for this title, appear on
page 58. Started only one month
before the Montreux Festival it won
a gold award. The graphic designer's
aim was 'A continuous
transformation between different
objects as exquisite and seamless as
conjuring'.
Graphic Designer/Graham Kern
Computer programmer/Gareth
Edwards of CAL Video Graphics

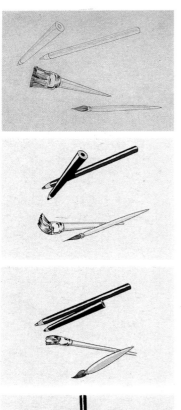

On the left is the first of fourteen pencil drawings of brushes flying through the air. These were line-tested to check the movement and then each was traced and painted on to cels, perfectly registered on a peg bar, before being shot twice (Double-framed), on a rostrum camera. These 28 cels made just a fraction over one second of animation.

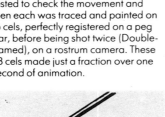

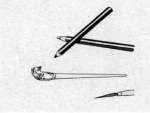

Twenty-four frames per second!

This page may give some idea of the amount of work required to produce effective movement and show the labour-intensive nature of conventional hand animation.
Form, shading and detail have to be limited because whatever is put on to one cel has to be repeated.
The graphic designer in television has often to produce similar work almost single-handed while the professional animation studios use many people to produce each sequence.
A few seconds for a childrens' programme, called 'Splash!' had over 42 drawings to form the face of a fashion model with a series of strokes made from brushes and eyebrow pencils.
Level 'A', shown above, presented the brushes and level 'B' drew the

face as 'C' made the marks from the brushes. Part of the 'dope sheet' – the instructions given to the rostrum operator – is shown opposite.
When completed this tiny section of the total 35 seconds was reduced to a 'bubble' using digital effects (Mirage and ADO) and floated across the screen. This can be seen in the sequence printed on page 91.

'Splash!'/Thames Television/UK
Graphic Designer/Barry O'Riordan
Animation/Reg Lodge
Rostrum camera/Norman Hunt and
Vic Cummings

The film rostrum camera

Rostrum film cameras like the Oxberry (shown right) have served graphic designers since the first days of television transmission. Some of the finest television graphic work, from the past and current output, has been achieved by the imaginative use of this extraordinarily versatile tool. The potential to mix and fade images at various speeds – to superimpose images, drawn and photographic – and to combine these frame-by-frame to a soundtrack has been an essential part of designing for television, and many examples of this craft are shown in this book. The challenge of the computer-aided graphics has not made the film system obsolete. The combination of computer-control and the motorised film rostrum, made about ten years ago, created new effects – some are shown in the following pages.

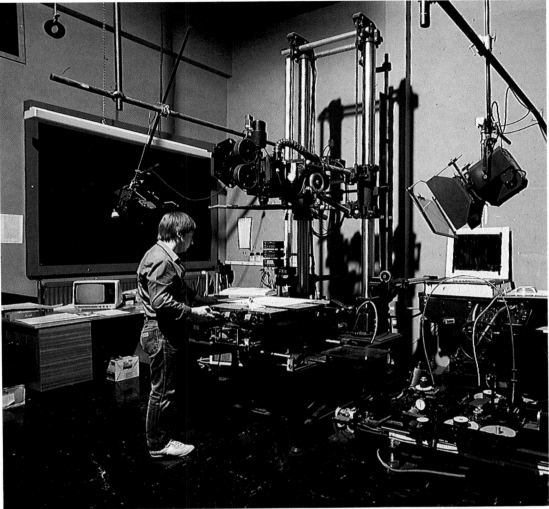

Photograph/Mike Cumpper

Below

'Hey Good lookin!'/London Weekend Television/UK

Multiple level cel work allowed a great deal of complex action in this hand-drawn rostrum camera title. Heads moved, fingers snapped, sparks flew – all in the same frame. The programme on design style was produced for Channel 4.
Graphic Designer/Pat Gavin
Animation/Pat Gavin and Jerry Hibbert

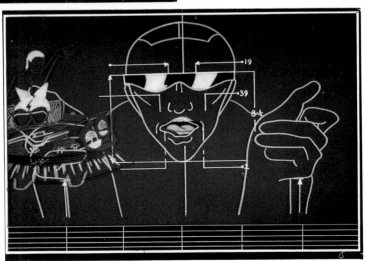

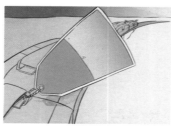
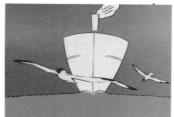

Saturday Night People'/London Weekend Television/1978/UK

Graphic Designer/Colin Robinson of Robinson Lambie-Nairn

Illustrator/Bob Norrington
Rostrum camera/Trevor Bond Associates

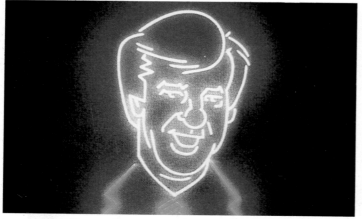
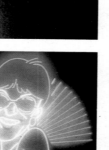
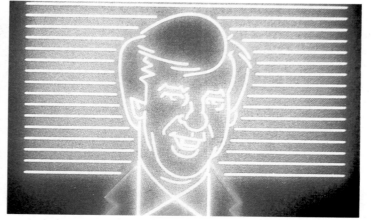

Cel animation

Transparent cels allow changes in drawings to be made and registered one on top of another on varying backgrounds to create the illusion of movement. In this case they are top-lit. This traditional and most basic technique of film animation was used in the fast-flowing title sequence for a holiday travel series shown opposite. There were over 700 cels in the 32 seconds.

'Wish you were here . . .?/Thames Television/ UK (left)
Design and Animation/Barry O'Riordan
Production services/BM Animation
Rostrum camera/Norman Hunt Jnr.

Top and back-lit animation

These two methods can be combined to create more movement and a greater feeling of depth. In 'Spectrum' the printed circuit artwork was top-lit and the camera was constantly zooming away from this plane while the back-lit line drawings were superimposed on another pass of the camera. The sound track was electronic music and the title provided a lively start to a children's programme on past and present scientific progress.

'Spectrum'/Thames Television/UK
Graphic Designer/Mick Mannveille
Rostrum Camera/Peter Goodwin

Rotascope

Live-action film or video can be traced from a projection screen, or monitor, one frame at a time and the drawings produced can be the source of reference for hand-animation or used as the basis for some stylised sequence. Chaplin's movements were viewed, analysed and choreographed into one swiftly flowing 'dance' for this opening title.

'Unknown Chaplin'/Thames Television/UK
Designer/Barry O'Riordan
Rostrum camera/Peter Goodwin

Hand-painting a cel on animation disk where the cels are pegged for accurate registration.

Back-lit cels

When painted cels, or more frequently photographic negatives (Kodaliths), are placed on a light-box on the rostrum table two effects can be achieved. The colours can be easily changed by coloured cels, or with filters on the camera, and the amount of light can be varied from very low to flaring which imitates the halation caused by neon tubes. This type of film animation was used for the late night chat show titles seen on the left.

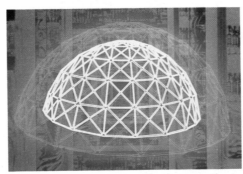

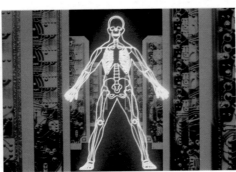

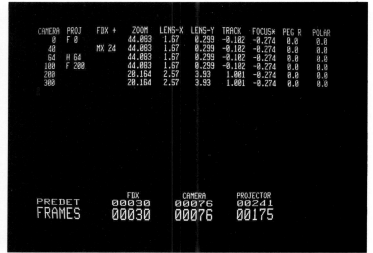

CAMERA	PROJ	FDX +	ZOOM	LENS-X	LENS-Y	TRACK	FOCUS*	PEG R	POLAR
0	F 0		44.083	1.67	0.299	-0.102	-0.274	0.0	0.0
40		MX 24	44.083	1.67	0.299	-0.102	-0.274	0.0	0.0
64	H 64		44.083	1.67	0.299	-0.102	-0.274	0.0	0.0
100	F 200		44.083	1.67	0.299	-0.102	-0.274	0.0	0.0
200			20.164	2.57	3.93	1.001	-0.274	0.0	0.0
300			20.164	2.57	3.93	1.001	-0.274	0.0	0.0

	FDX	CAMERA	PROJECTOR
PREDET	00030	00076	00241
FRAMES	00030	00076	00175

Many film rostrum cameras, similar to the one shown on page 105, have over the past ten years benefited from computer-control. All the functions have been motorised, and via a keyboard, (above left) every movement of the rostrum bed, the camera exposure, the zoom on the main column, and even the movements of an associated aerial image system, have been able to be programmed to carry out panning and tracking to a complexity well beyond the power of manual operation. A VDU (above right) presents and records the sequence of the operations using a very direct language already familiar to rostrum camera operators. Once programmed, the speed of filming can approach the real time of the final projected sequence. Computer-control allowed the development of a number of techniques, and the two most commonly used have been 'streak-timing' and 'slit-scan'.

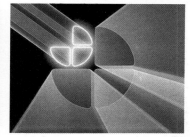

Slit-scan
The type of distortion made possible by computer-control of the rostrum camera is shown above. It is from a title made by Bernard Lodge who contributed a great deal to the extension of this form of film animation. The example below is from 'Young Musician of the Year'.

Streak-timing
This uses the blur of light during a long exposure on one frame of film. The examples are (left) 'The Crystal Cube' (BBC), and (above) from TV Globo – 'Esporte Espetacular'.

'Young Musician of the Year/BBC/UK
Both the film animation techniques described opposite, and in the main text, are employed in this title. The penultimate frame shown here inspired the trophy design for the series.
Graphic Designer/Pauline Carter

'Miss World 1985'/Thames Television/UK
The theme and the graphic title of this programme are always related. In 1985 the idea was based on the song 'Luck be a Lady Tonight' and 'streak-timing' was used to present the title lettering.
Graphic Designer/Julia Stone
Film animation/Cel Animation Limited

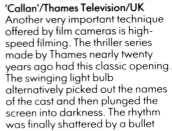

'Callan'/Thames Television/UK
Another very important technique offered by film cameras is high-speed filming. The thriller series made by Thames nearly twenty years ago had this classic opening. The swinging light bulb alternatively picked out the names of the cast and then plunged the screen into darkness. The rhythm was finally shattered by a bullet and caught by filming at 128 frames per second. Video systems are now available but with a present maximum of only three times the normal recording speed, film has the clear advantage.
Graphic Designer/Ian Kestle
Film camera/Vic Cummings of Caravel

The illustrations on this page are examples of model animation for television graphic design where film cameras have been used. They reveal some of the wide range of applications modelmakers and graphic designers can employ.

The two photographs (right), taken at Damson Studios, show a title sequence being shot using a moveable model. The film camera is mounted on the mobile dolly. The stack of 'books' was carefully made so that their position could be controlled very precisely as the action required. The camera, an Arriflex, can shoot one frame at a time and the dolly was guided by tracks on the floor of the studio, allowing movements to be repeated. These photographs were taken during the production of a title for Thames Television's Light Entertainment series 'Executive Stress'. The graphic designer was Rob Page.

Model animation for 'The Kenny Everett Show'
The Kenny Everett production team involved graphic designers very strongly from the early days of transfering from radio to television.

The 'electric' feeling for 'The Kenny Everett All Electric Show' was gained by projecting moving coloured light on to the highly-reflective material and cut-out polystyrene letters sprayed with metallic paint.
Graphic Designer/Jim Gask

A model 'proscenium', inspired by 'Lord Thames Video Vault' jokes, was about three feet by two feet and became the setting for a number of model effects. The construction was DIY plastic piping painted white, and lit with coloured gells.
Graphic Designer/John Stamp
Modelmaker/Freeborns

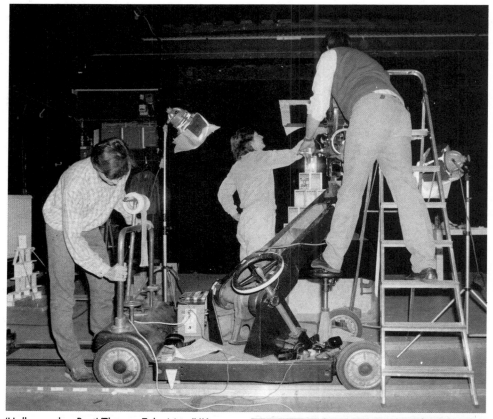

'Hollywood or Bust'/Thames Television/UK
A clear perspex sphere was filmed whilst rotating on a motorised rig. The sphere was then sprayed matt black round the cut-out lettering applied. This allowed perfect registration on the second shoot. The coloured drapes provided enough reflected colour and pattern to heighten the effect.
Graphic Designer/Ruth Bribram
Modelmaker/Stephen Greenfield
Modelmakers
Model Animation/Damson Studios

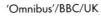

'Omnibus'/BBC/UK
An elaborate and large-scale model was
made for this title. The control of the shoot
depended on the skills equal to those of
Lighting Director over the whole 'set'. The
view of the back of the model shows a detail
of some of the bulbs and wiring required.
Graphic Designer/Darrell Pockett
Modelmaker/Stephen Greenfield

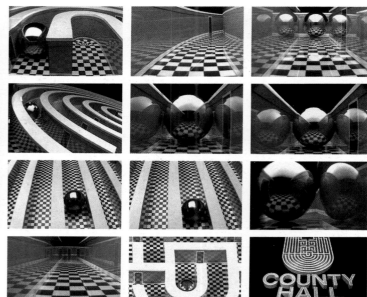

'County Hall'/BBC/UK
Above is a photograph of the model which was constructed to suggest a maze and the sometimes impenetrable corridors of local authority power. The moving ball was filmed by Oxford Scientific Films who specialise they say 'in getting cameras into unusual positions, high-speed and micro-cinemaphotography'.
Graphic Designer/Michael Graham-Smith. Modelmaker/
Stephen Greenfield Modelmakers

'Private Lives'/BBC/UK
Within the wooden head were many segments with more models. The conversation in this chat show was sparked off by a song, a place, memory – the bric-a-brac that is inside everyone's head. The show was hosted by the actress Maria Aitken.
Graphic Designer/Michael Graham-Smith. Modelmaker/John Friedlander
Cameraman/Douglas Adamson

'Tinker Tailor Soldier Spy'/BBC/UK
No special model-making was required for the wooden dolls – they were purchased and repainted to alter their facial expressions – but a metal armature was made to control the precise movements required for filming this memorable title for John Le Carre's spy series.
Graphic Designer/Douglas Burd
Modelmaker/Percy Packman

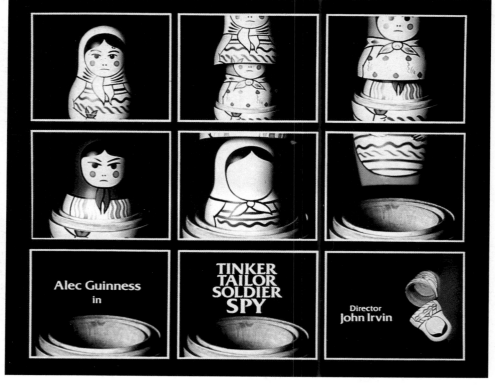

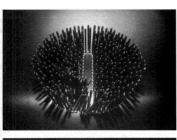

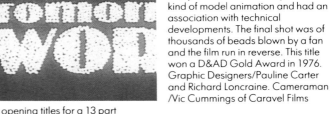

'Tomorrow's World'/BBC/UK
Every letter was formed by a different kind of model animation and had an association with technical developments. The final shot was of thousands of beads blown by a fan and the film run in reverse. This title won a D&AD Gold Award in 1976. Graphic Designers/Pauline Carter and Richard Loncraine. Cameraman /Vic Cummings of Caravel Films

The opening titles for a 13 part documentary on the early history of Hollywood used a model animation camera for a 'multi-plane glass shot'. By mounting cut-out photographs, at various scales, an extremely long zoom-out was achieved. The camera started on a close-up, on a field-size of only 2 inches, of Clara Bow at the centre of the set and slowly pulled away to reveal more and more of the studio and surrounding lot. Black-and-white was used in homage to the period.

The illustrations (below right) show the layers of glass and (left) the model animation camera and its mounting.
'Hollywood'/Thames Television/UK
Graphic Designer/Barry O'Riordan
Animation/The Film Company

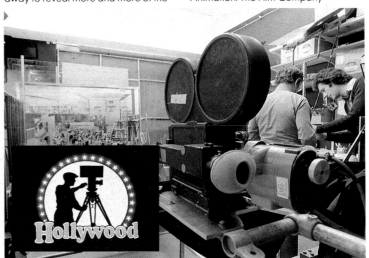

Computer-control has been applied to model animation to give more precision and greater freedom of movement than the older manual systems. Sequences can be stored in memory and repeated with totally controlled registration for multiple exposures, test and re-runs.

The rig shown here (1) was custom-built at The Moving Picture Company and developed by Peter Truckle. The very strong overhead tracks guide either a video or a film camera, with single frame recording capability, at a thousandth of an inch in any direction.

A snorkel lens demonstrated here (2 & 4) allows the camera to rove in all axes at any apparent speed. From a single console in the studio the model can be simultaneously moved a thousandth on an inch at a time on a separate rig under computer-control with the same potential of exact repeatability. An example shown below (3) is from a promotion sequence from Channel 4 designed by Bob English in 1984 where the back-lit model of a city was nearly eight feet across and the camera slowly zoomed into a targeted sight from what appeared to be the height of several hundred feet.

Film back-projection, colour separation overlay or any combination of these and similar techniques, can be used to extend the illusions and produce complex special effects on film and video tape.

'Soldiers'/BBC/UK
In the BBC title sequence for 'Soldiers' (right) the camera performed a horizontal flight above the ranks of the model figures with breathtaking skill over a distance of about ten feet.
Graphic Designer/Liz Friedman

'Equinox'/Channel 4/UK
A series on science and technology. A model was designed and built to deflect a low-powered laser beam, through a series of mirrors, eventually creating a face which mouths the word 'Equinox'. Unusually, the title did not appear on the screen. The motion control rig was used to repeat identically some of the camera moves, as a double exposure on film was necessary to achieve the desired balance between laser light strength and background illumination. Video effects were used to create a face inside the beam of light.
Design and Direction/Bob English, English Markell Pockett
Models/Steve Wilsher, Creating Effects
Lighting cameraman/Pete Truckle

4

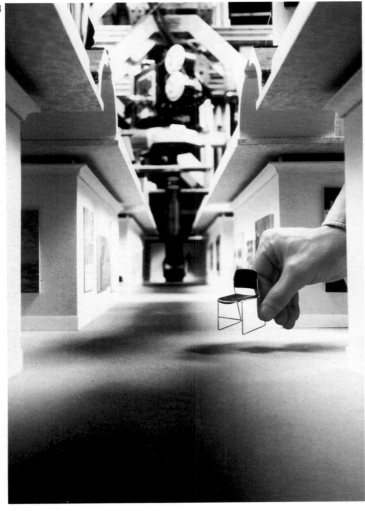

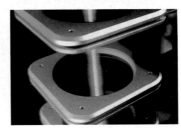
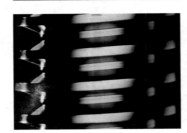

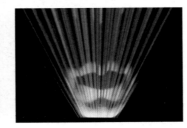

TV Globo in Brazil carried out a very large-scale model animation for a drama series title. Two separate constructions were filmed in front of a large back-projection screen using cloud effects. The first model set up a modern city skyline and the second at first appeared to be a distant view of the same city but as the camera angle changed to an aerial shot the mass of buildings revealed a huge portrait of a man's face.

Selva de Pedra TV Globo 1986
Art Direction/Hans Donner
Design/Hans Donner/Gustavo Garnier/
Nilton Nunes and Richardo Nouenber

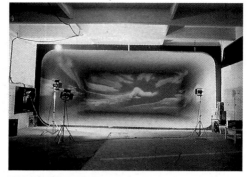

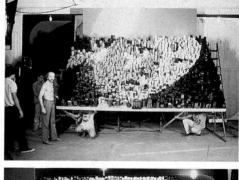

Recording images with video animation, rather than film, has been an objective throughout television history.

The two main advantages of a video system are instant re-play as opposed to waiting for exposed film to be processed, and the difference in picture quality between film and electronic images.

Until there were stills stores with reliable and fast methods of storing and re-playing single frames, the construction an electronic rostrum camera was impractical.

The BBC have created the system, shown here, where graphic designers can now prepare a large proportion of their programme animations in video. An account of the development is given in Section 6c.

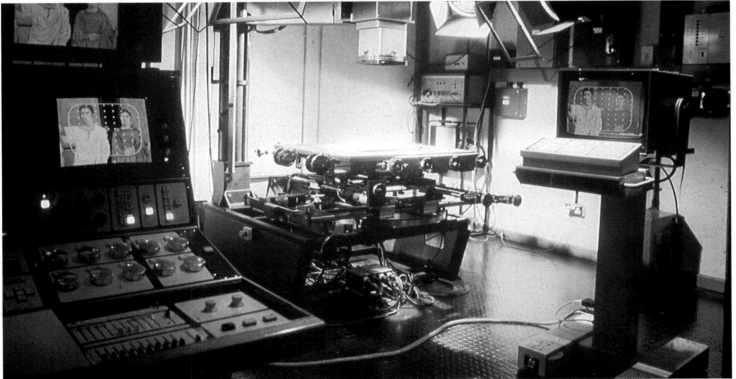

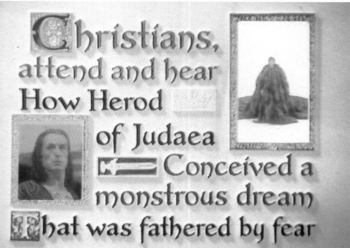

Keying pictures from two, or more, video sources opens-up limitless possibilities for combining pictures, photographs, models – in fact any image – with actors at almost any scale.

The background to these systems is described in the main text while the examples on these pages show some ways Chromakey, and Ultimatte, (ITV terms) and CSO – Colour Separation Overlay (the BBC designation) have been applied to programme making.

Broadly, the graphic designer and the set designer can use their skills and imagination, by working closely together, to remove the need for scenery. The intention can be to create effects otherwise impossible, or to save costs! All the photographs on this spread come from a production of 'L'Enfance du Christ' by Berlioz where the entire concept of the programme relied on Ultimatte. In the shot of the studio (1) a very large area, and the cyclorama, are covered in blue material. This colour is keyed-out and only the characters and props were recorded. The television camera is being aligned with large 'registration marks' to fit marks on the 'graphic' (5), a picture only about 20 x 30 inches, being recorded by a video rostrum camera. All the pictures were prepared by the graphic designer. Below (4) the actors, and the sheep, are in the studio and there is an off-screen shot (6) where the combination was achieved. The perspective was controlled to allow them to appear to move away under the trees.

Multiple video inserts are seen in (2), where there are a modelled outer frame with four keyed images and in a central scene three more Ultimatte levels, (3) shows the method used to present the many captions, or 'subtitles', in the work.

'L'Enfance du Christ'/Thames Television/UK
Graphic Designer/Barry O'Riordan
Production Designer/Peter Le Page

5

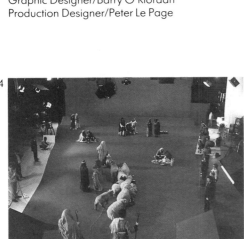

4

6

Colour separation is regularly applied to childrens' programmes on the dual grounds of economy and achieving fantasy effects.

Many illustrations, like the one above (1), were made to create a setting – to add or remove clouds, trees, and 'houses' in 'Wobbly Land'. The small off-screen shots (2 and 3) show the performers in-action in a sequence where a third level of Chromakey was used to make the jelly illustration (4) wobbly, using the vision mixing desk.

To provide as many illustrations as possible, for another programme in the same series, a model was made of a toy fort. This was photographed from many angles and combined with suitable drawn backgrounds (5).

The simple drawing (6) was enough to make a setting where the characters appeared in the mirrors, Further off-screen shots (7 and 8) suggest how few props and furniture are required to achieve many scene changes.
'Rod, Jane and Freddie'/Thames Television/UK
Graphic designer and illustrator/Rob Page

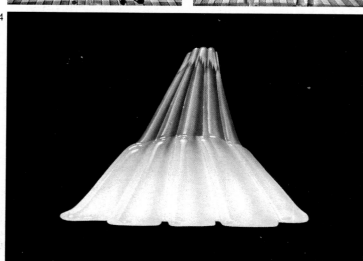

The BBC produced a light-hearted series bringing the Second World War newspaper strip-cartoon character 'Jane' to the television screen. The production combined the fast output of hundreds of Paintbox drawings with the colour separation overlay system to add the action of the cast. The backgrounds were in simple colours while the performers were recorded in black-and-white to emphasize the newsprint style.

'Jane'/BBC/UK
Graphic Designer/Graham McCullum

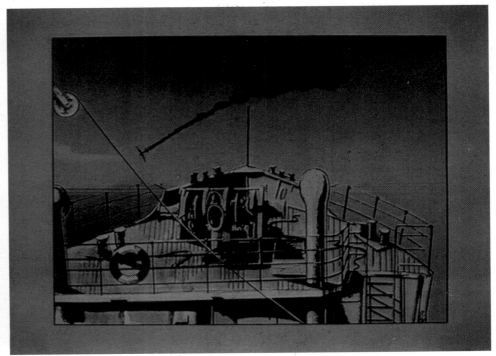

Digital effects equipment

Manipulating still and moving images in television production has advanced with the progress of digital technology. Bending, twisting, and turning – both moving and still pictures – is a facility now commonplace. Perhaps too frequently used and often at odds with the essence of graphic design which is to plan and control images with a clear purpose.

Effects for the sake of movement alone are distracting and very quickly repetitive. There are now signs that the worst period is over; graphic designers have learned more about the machines; and there is a better relationship between technicians and designers.

Some devices, now sitting more comfortably in the new environment, are shown here.

Stills stores

For a very long time graphic designers in television stored all their artwork in the same cumbersome ways as graphic designers working in print and advertising. Piles of captions, and drawings, maps and diagrams prepared on artboard were filed in plan chests.

Now there are electronic stills stores (ESS) where thousands of single frame images can be stored in digital form on disks, indexed, and rapidly accessed from libraries of up to half a million pictures.

Logica's Digital Library System-Gallery 2000, shown here, stores every picture as a numbered and reduced size image – 30 of these can be viewed at one time, (above) while more frames can be scrolled through instanaeously. Below left, is one of the indexes, and (right) an operator at the keyboard.

Easy access, compact storage and the ability to call up a precise undamaged 'original' within seconds has revolutionised the television graphic studio and been the basis of computer-aided animation and much digital effects equipment.

Mirage

Quantel describe Mirage as 'a true 3D picture manipulator in 3D space'. At its debut – USA in 1982 – its wizardry made flat artwork into globes and cylinders, it produced instant 'page turns'; pictures exploded or turned into multiple moving geometric forms. Production companies purchased it very quickly – in its NTSC form it went first to Broadway Video, New York in April 1983 – but the brilliant, though too obvious, effects put graphic designers off and there was a period before its place in post production was consolidated. It was then realised Mirage could be programmed to achieve the 'impossible job' and became the delight of those who now use it.

Ampex ADO

ADO stands for 'Ampex digital optics' and this device gives space and time manipulation. Pictures can expand from infinity to 200 times normal, or they can rotate in any direction or combination of directions. Full perspective translates from 2D to 3D and all programmed moves are repeatable or they can be stored in the memory.

Quantel's Encore

Encore is described by Quantel as 'a 2D picture manipulator in apparent 3D space' with a range of effects including zoom, compression, positioning, rotation and perspective. It allows up to 14 pictures from separate units to be displayed on one screen. The frame above shows a mosaic effect on a 'flying' picture and on the right solarisation.

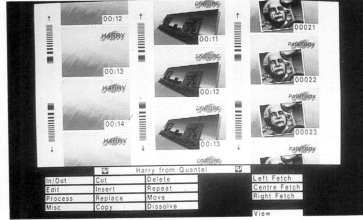

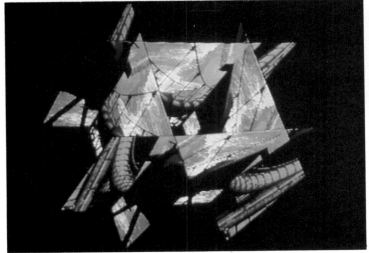

Quantel's Harry

Television effects have been described as anything in a programme *not* produced by basic shooting or normal editing. Acres of space have been devoted in the technical video and design press to define 'Harry'. Is it an editing unit or a special effects device? In its basic form it can record up to 2000 frames, that is 80 seconds of real time video.

Each frame can be accessed in totally random order and then dissolved, frozen, or keyed with any other frame and then replayed in any order. Linked to a Paintbox it can retouch multiple frames or start from scratch to create any type of animation. As a 'digital studio in a box, and an animation aid of incredible flexibility' Harry's usefulness to graphic designers has only just been grasped.

Live-action material, either especially shot or taken from library footage has been one of the most pervasive elements of opening title sequences.

'Match of the Day'/BBC/UK
An example from 1967 where live-action film was integrated into a moving football rattle using mattes and masters.
Graphic Designer/Colin Cheesman

'Match of the Day'/BBC1/UK
Fast clips of play were intercut with further live shots of the grandstand crowd. They played their part by forming massive picture mosaics. This was very influential when it was made in 1975.
Graphic Designer/Pauline Carter

'We Wish You Peace'/CBS/USA
Storyboard and shots from a 10-second Christmas holiday message for the Entertainment Division of CBS made in 1985. Its making is described on page 30.
Graphic Designer/John C LePrevost
Live-action production/Apogee Incorporated

'Good Morning Britain'/TVam/UK
Some of the problems and excitement of making this succession of brief live-action shots to launch the ITV breakfast programme in 1982 are described on page 21.
Graphic Designer/Ethan Ames

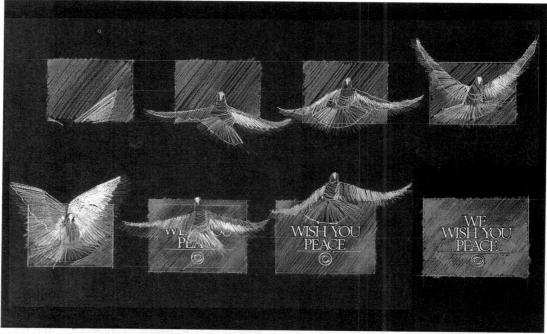

Photograph/Vincent McLeary

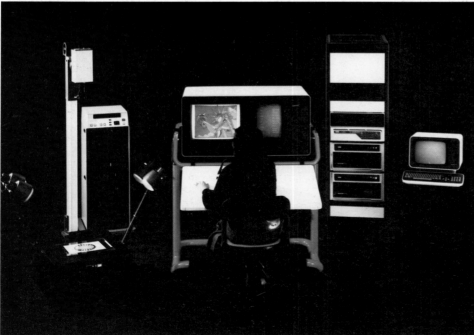

(Below) left is a photograph of the Ampex digital paint system called 'AVA'. It was the most advanced of its kind in the early 1980s.

In barely five years, graphic designers throughout the entire television industry have become familiar with a new world where they produce much of their design work directly in the video medium on digital paint systems, similar to the version here.

The electronic stylus and the keyboard are two options for putting information on-screen and both can be seen in this photograph. The video camera (below) enables any photograph, drawing, map, engraving or even three-dimensional objects, to be viewed on the monitor and instantly digitized into the system's memory.

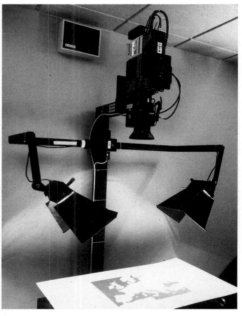

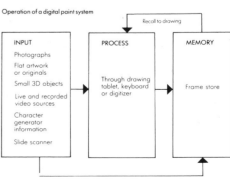

Operation of a digital paint system

INPUT	PROCESS	MEMORY
Photographs		
Flat artwork or originals	Through drawing tablet, keyboard or digitizer	Frame store
Small 3D objects		
Live and recorded video sources		
Character generator information		
Slide scanner		

Recall to drawing

Direct feed from video or character generator

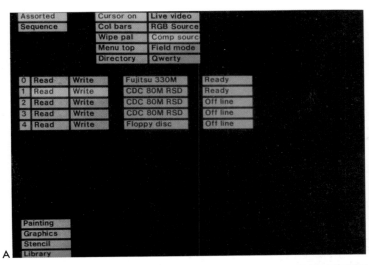

	Assorted		Cursor on	Live video
	Sequence		Col bars	RGB Source
			Wipe pal	Comp sourc
			Menu top	Field mode
			Directory	Qwerty

0	Read	Write	Fujitsu 330M	Ready
1	Read	Write	CDC 80M RSD	Ready
2	Read	Write	CDC 80M RSD	Off line
3	Read	Write	CDC 80M RSD	Off line
4	Read	Write	Floppy disc	Off line

Painting
Graphics
Stencil
Library

A

B

Painting	Find		Fetch
Graphics	Save		Delete
Stencil	Load	Down	
Library	Dump	Titles	
Pasteup	Picture	Browse	
		Archive	

C

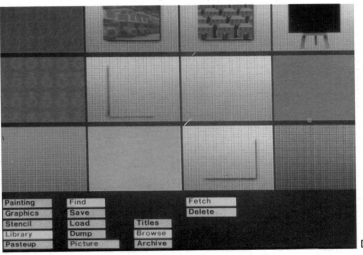

Painting	Find		Fetch
Graphics	Save		Delete
Stencil	Load	Titles	
Library	Dump	Browse	
Pasteup	Picture	Archive	

D

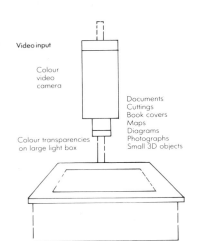

Video input

Colour
video
camera

Documents
Cuttings
Book covers
Maps
Diagrams
Photographs
Small 3D objects

Colour transparencies
on large light box

The list of operations many digital paint systems provide is now quite extraordinary. Any image in full-colour, once in the memory, can be enlarged, or reduced, repeated at will, reversed left to right – or positive to negative, retouched in detail, made into a stencil to be 'cut out' and 'pasted-up' anywhere on screen. Every medium known to artists can be simulated so well that the image on-screen can imitate watercolour, pastel, airbrush work and the impasto of oil colour.

The modes in which the machine operates are selected from the 'menus' displayed on the monitor. Illustration 'A' shows some of the many commands which appear. When the stylus is pressed on the command required the system is instantly ready to carry out that function. Select the colour of the background – it appears. Select the colour for the 'pen' – it is ready to draw. Change to 'airbrush' (fine or coarse) and commence.

Illustration (B) shows a section of the colour palette available for mixing the precise colour you wish to use. The range is a mere 16 million. The picture on the monitor shown here is from a colour slide and any part of it could be retouched to remove, or increase the clouds, or draw in any object desired.

(C) presents a sub-section of the menu which is recalled to change the mode while work on an image is in progress. The final illustration (D), reveals the browse mode where partially, or fully-completed, elements of the design stored in the memory can be reviewed.

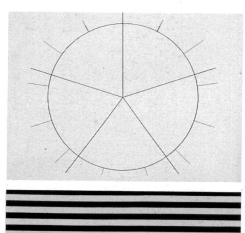
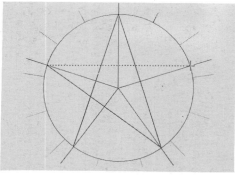

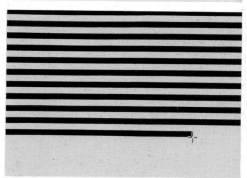

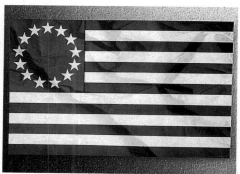

The speed and the directness of preparing images for television using the digital paint system are now legendary. Above is a 'test piece' prepared for this book by Richard Bain of Quantel on a Paintbox to demonstrate the measured and geometric aids to drawing contained within the Paintbox Pro4 software.

Only 35 minutes were required to produce the flag and this could be reduced or repeated many times, within seconds. To prepare artwork using paper, gouache and airbrush to the same standard would take hours rather than minutes and repetition and reduction involve photo-processing.

Early models did not have type within the memory. Now most electronic paint systems do offer lettering. On the right are examples of ways a basic fount can be amended with shadow, relief and outline by using the Paintbox graphic techniques.

One of the earliest networks to harness an electronic drawing unit to its news programme was 'CBS News' in New York in 1981. These are samples from the Ampex AVA machine of that period.
Art Director/Ned Steinberg

On the opposite page is a collection of end uses to which electronic paint systems are applied in television graphics to present still images on-screen.

1

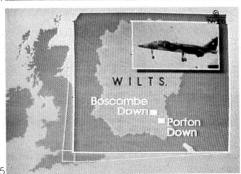

4

INTERNATIONAL DRESSAGE

4

2

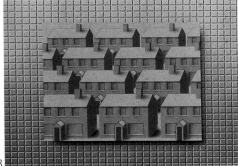

WILTS.

Boscombe Down

Porton Down

5

3

1985

6

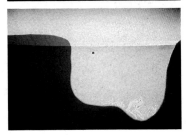

A 'chalk' drawing (1) Graham McCallum, of the BBC, on Logica's 'Flair' showed very early in the development of the digital drawing devices just how subtle and light the touch could be. Channel 4 is one of the many networks to produce promotion captions (2) with electronic drawing. The finished detail and precision of these items (3) from 'Thames News' could not be achieved so quickly. Map outlines can be stored and speedily up-dated with new information as seen in these samples from TVam, (4 & 5), very useful in a story running over several days. Even the station copyright caption (6) has been stored and amended year by year, 'Add an egg!'

Information graphics, where multiple drawings are required to develop a story-line or explain a process, can be served very well by the ability of the digital paint system to store, repeat and change a single image, add further stages, then replay them sequentially. The 'Quantum' sequence demonstrates this.

More recently these systems have been a vital part of producing images for animated sequences in conjunction with other computer hardware and developing animation via Quantel's oddly named 'Harry'. The retouching power has extended set design in the way that 'glass shots' are used.

'Quantum'/ABC Television/Australia
'Quantum' is a science programme and these frames illustrate the formation of fossils in the area of North Queensland.
Graphic Designer/Ann Connor

Computers have gained great control over the television screen. They can produce a single frame in full colour of almost any complexity. They can also make multiple frames to replace, in computer terms, the cels used in conventional animation. How this can be achieved is described here in a very simple form.

The 'ray' of a cathode ray tube (CRT) illuminates only one spot of the picture at any one time, and the rapid scanning of a television picture in line form (raster) creates a steady image due to the persistence of vision. Another visual illusion creates each colour by using different brightnesses of red, green and blue light for each picture element (pixel). A typical full-screen television image totals 300,000 pixels from about 500 lines divided into at least 600 elements, and each pixel can be represented by three 8-bit digits. These numbers, or 'bytes', correspond to the RGB intensities of each pixel. Therefore to control a tv picture a computer has to store about one million bytes in a frame store. Such a store is not much use unless we can easily address and manipulate the contents. That is the function of the central processing unit (CPU).

Any sort of realtime animation requires the picture store to be updated 25 times every second, and 25 million functions a second is close to the upper limit at present but simpler one step at a time operations are quite easy for the CPU to perform. For example any two pixels can be linked (1) to create a line. More points, similarly activated, can form a shape (2). Using mulitple polygons any shape can be created and stored, (3), as a wireframe figure. A transparent cube? Using more information, where edges but not surfaces are rendered, lines which can not be seen from a chosen viewpoint can be concealed (4). This is called a 'hidden line system' and helps to create three dimensional form. Once established a shape can be enlarged, reduced, or repeated ad infinitum (5). Further software can control perspective (6) and make in-betweens for animation. (7) and render surfaces as solids where the effects of varied sources of light can be manipulated as required when movement takes place. Still more information computes shadows (8) and reflections, (9) colour, texture and reflective or matt surfaces to simulate every type of material – from transparent perspex to coarse grain stone. An image like the frame shown in (10) takes those million pixels, and 25 frames a second to produce smooth, 3D, fully-rendered animation. All you need is a little box of tricks to deal with 25 million functions a second! Even the most powerful computer can take several hours to build up such a complex single frame so the process is very costly.

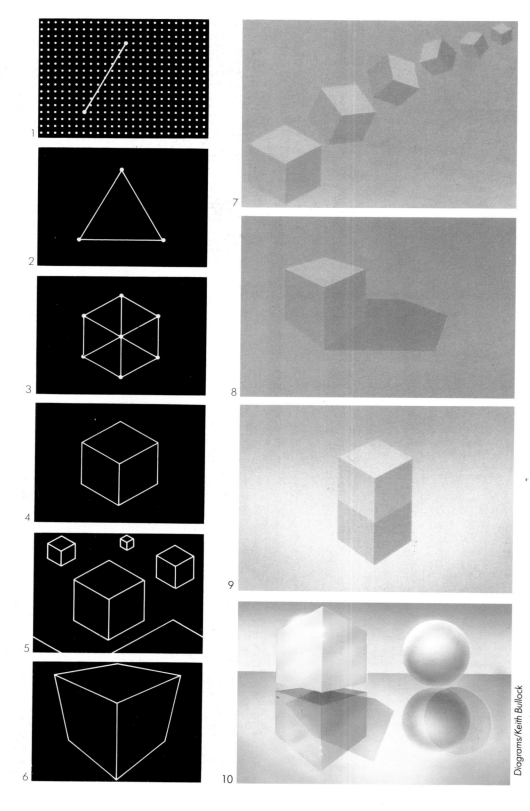

Diagrams/Keith Bullock

The Renaissance woodcuts are a reminder of the long and vital relationship between art and science in picture-making. It is also a tribute to the immensely important part mathematics has played in all modern computer-aided design. Graphic designers do not have to understand calculus or advanced maths; although they will benefit from some knowledge of the computer programmers world. They must know what the computer and the programmers are able to achieve and design – most of the time – within these parameters.

(Above right) Light and shadow control in motion using three dimensional computer-aided animation is demonstrated in this single frame from the station ident. The sequence was designed and produced by Robinson, Lambie Nairn for Scottish Television. Computer animation/Electronic Arts.

(Below right) Complex 'ray tracing' enabled the American computer animation company Cranston Csuri Productions of Columbus Ohio to produce this extra-extraordinary effect in 1983 where two 'prisms' turned slowly in front of a picture and refract in a strangely faithful manner the scene behind them. 'Ray tracing' uses algorithms which are sets of complex rules for solving problems in a finite number of operations and render surfaces as solids where the effects of various sources of light can be manipulated as required when movement takes place.

'Music in Time'/Polytel Film Limited/ Lanseer Films
Before current advanced technology arrived, to fully animate almost any given storyboard, graphic designers tried computer-aids which still required a large amount of hand labour. The metronome, violins and the Albert Hall were all constructed with in-betweens evolved on a computer then drawn via a line-plotter at the Middlesex Polytechnic. The drawings were transferred to cels for handpainting and the busts were model film animation.
Graphic Design and Production/ Robinson Lambie-Nairn

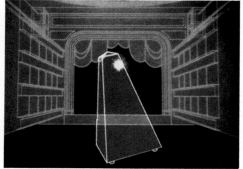

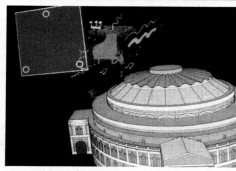

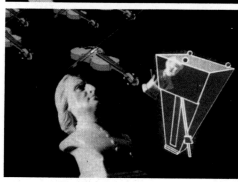

'CBS Late Night/USA
A highly-finished storyboard for this twenty-second promotion animation is on page 5
Creative Director/John C. Le Prevost
Designer/Mark Hensley
Animator/Roger Gould of Pacific Data Images

'Star Wars'/ABC/New York
A two minute sequence illustrating the workings of SDI 'star wars'. For ABC Television in New York. Part of a large networked documentary called 'The Fire Unleashed' covering all aspects of nuclear energy.
Design and Direction/Richard Markell, English Markell Pockett Programme Company/ABC Television (New York) Computer animation/Computer FX

'Sporting Triangle'/Central TV/UK
Only three years ago Electronic Arts were asked by the graphic designer working on the BBC Olympic Games title – 'Can you do shadows?'. Computer art has moved so fast that people do not have to ask very much now as shown by the realistic leather and stitching on the cricket ball. Five sports emerged from the rolling dice and the 26 seconds cost £16,000. Electronic Arts technical description: 'Foreground image-choreography created by using Wavefront previews on the Iris. Colour rendering by Wavefront on Pyrammid. Background on the VAX rendered on to film. Glow effects by in-house software.'
Graphic Designer/Geoff Pearson
Animators/Lai Chan, John Roberts-Cox, Jilly Knight and Nick Glazzard

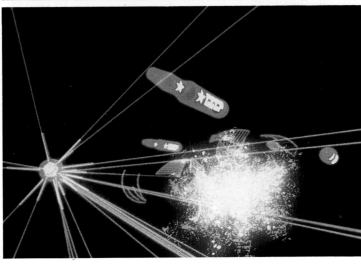

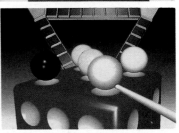

The Innovators/Airshow International
Title sequence for a series to be transmitted in America featuring the best creative designers from many disciplines including fashion, industrial architecture and graphics. A complex series of images personify the breadth of the series were linked. As we find each image the 'thought' activates a change from conception to reality.

An account of this computer-aided animation is on page 33.
Graphic Designer/Darrell Pockett, English Markell Pockett
Computer animation/Gareth Edwards of Cal Videographics

'Man and Music'/Granada Television for Channel 4/UK
The director who commissioned this challenged the graphic designer to produce the title without any visual reference to musical instruments, notation, or musicians. How the designer developed this solution, which was set to 16 seconds of Bach's D Minor Harpsicord Concerto, is described on page 31. This animation is rendered in the vector lines of the computer rather than the solid raster shading.
Graphic Designer/Bernard Lodge
Computer animation/Derek Lowe of Computer FX

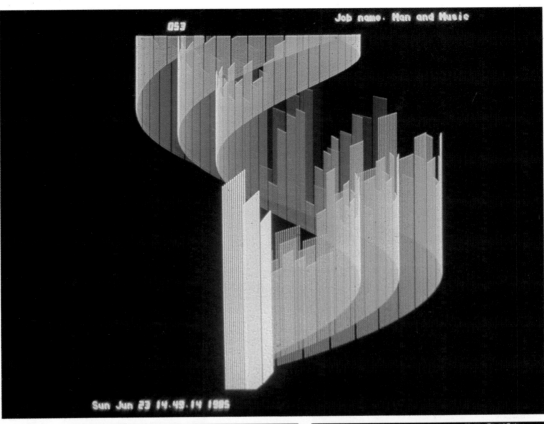

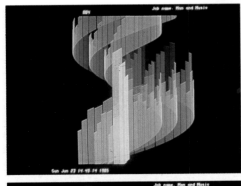

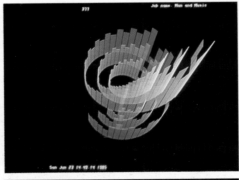

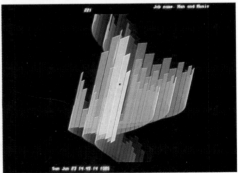

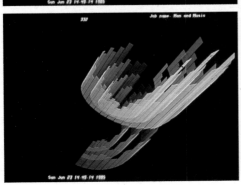

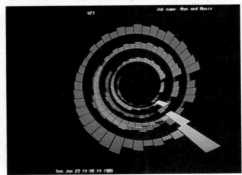

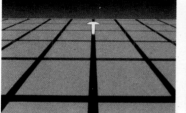

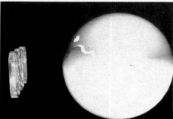

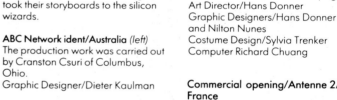

International boundaries were swiftly broken down by the application of space, flight and military computer science progress made in the USA to entertainment and TV graphic animation. Rapidly South American, European, and Australian designers took their storyboards to the silicon wizards.

ABC Network ident/Australia *(left)*
The production work was carried out by Cranston Csuri of Columbus, Ohio.
Graphic Designer/Dieter Kaulman

'ABC National'/Australia *(above)*
An animatic in wireframe was produced at XYZap in Australia and then sent to Digital Productions in the US for completion on a Cray computer.
Graphic Designer/Terry Dyer

'Fantastico'/TVGlobo/Brazil *(right)*
The Brazilians went to Pacific Data Images in San Francisco in 1983 to realise a sequence combining computer animation with keyed live-action dancers.
The production account is on page **23**.
Art Director/Hans Donner
Graphic Designers/Hans Donner and Nilton Nunes
Costume Design/Sylvia Trenker
Computer Richard Chuang

Commercial opening/Antenne 2/France
Some networks use a brief graphic animation to signal the end of programme material and the start of advertising.
Graphic Designer/Xavier Nicholas
Computer Production/Sogitec

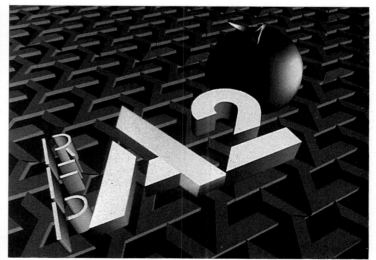

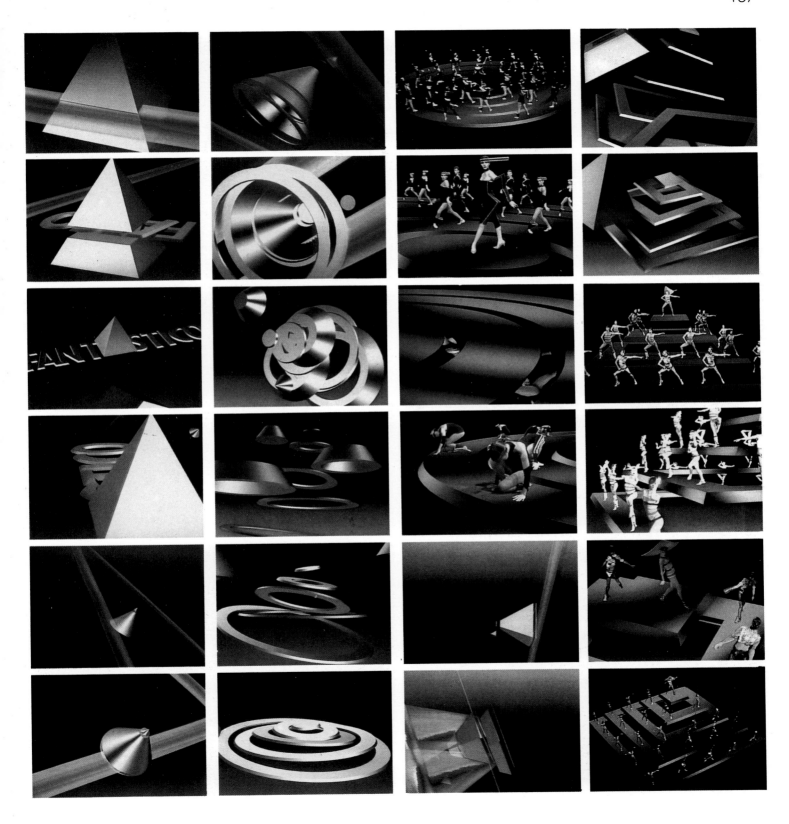

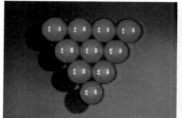

RAI/Station idents/Italy
Graphic designers have travelled to achieve the best results in computer animation, and have no national boundaries. TV Globo of Brazil used the New York Institute of Technology in the 1970s – British television designers have used American companies – RAI Uno, the major Italian government network, went to the French computer company, Sogitec. The high cost of this type of work has concentrated a lot of current effort on station idents and promotion design. Lower costs and more experience will see the spread to all areas of programme making.
Graphic Design/Video Italia

Computer animation/Sogitec

'Canadian Master Snooker'/CBC
Equipment has been produced to enable graphic designers to make computer animation sequences 'in-house'. The first were slow and the results did not compare with the work produced on the much more powerful processors in facility companies. The above sequences were designed and made in the graphic design department of CBC in Toronto on an Image 2 system. The unit is made by the Computer Graphics Laboratory at the New York Institute of Technology, and the time when every graphic designer will have access to desk-top 3D animation seems closer every week.

Graphic Designer/Gordon Morris

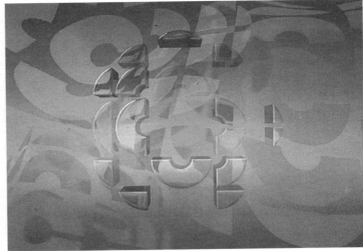

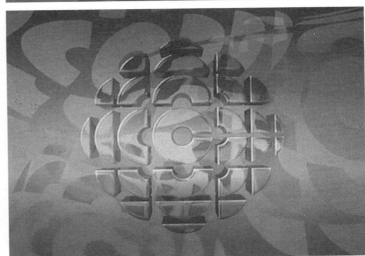

The Canadian Broadcasting Corporation used computer animation very early and their long established network symbol (right) was translated, by Cranston Csuri of Columbus Ohio, into transparent, perspex-like, reflecting forms in 1980. Here the advanced software available at Bo Gehring (they did the Channel 4 animation) produced a four second animated ident with colour variations to respond to different times of day as well as special occasions.

The large sequence shows the morning version. The first of the small shots the afternoon, the second a 'neutral' version, the third appears in 'prime time' and the fourth was used on the 50th anniversary.
Art Director/Gerard Bueche
Computer animation/Bo Gehring Aviation, Los Angeles

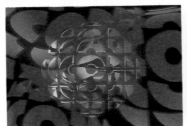

'Oh Boy' and 'The Church and Television'/ABC Television/1957c

These captions show hand-lettering was still being used for television twenty years after the first service was set-up in the UK! The style of lettering seems suitable for 'The Church and Television' but rather incongruous for the Jack Good's then shockingly wild teenage pop programme 'Oh Boy!'

Masseeley printed captions

After hand-lettering the great mass of on-screen lettering, required daily by every television network, was for many years, hand-set in metal type and proofed on a 'hot-press' similar to the Masseeley press in the photograph on the right. Bottom-of-frame captions, single captions and rollers, carrying the actors and technical credits as well as rundowns and end of parts were all set character-by-character in the five hundred year old system developed by Johann Guntenberg!

The Masseeley hand-press used metal type and the impression, usually on to black card to give the most acceptable image to the television screen, was made from white foil using a heated element. The advantage was a solid white letterform which was dry and ready for use, as a roller or single caption, immediately. No other device provided this and units served television stations from Reykjavik to Rio do Janeiro through at least three decades.

Rollers, which reveal large amounts of copy as they move up the screen, and dozens of captions could be produced. The process was very slow and the final printed image had to be transferred into a video signal, either by a television camera or intermediate photographic slides, and then a telecine slide scanner. The fastest time, for even a single line, from approved copy to transmission was about half-an-hour.

End credits

The examples opposite are from a BBC Shakespearean series where the graphic designer, Alan Jeapes, was able to use the improved colour transmission standards from 1968 onwards to reproduce the grain of the paper and even the impression of an old type-face biting into the surface. The filmed end credits for 'The English Garden' were designed by Bernard Allum of Thames Television. The BBC documentary 'Cash from Trash' displayed the production credits as still slides. Designer: Darrell Pockett.

From the bare necessity of listing the names of the cast and the technicians who produce the programme, the end credits began to be designed with a great deal of care to reflect the style of the opening title and to harmonise with the rest of the programme.

The form of the lettering can still range from founder's type, photosetting, hand-drawn lettering, transfer lettering – like Letraset, and even modelled letter forms and is limited only by the imagination and hopefully, the suitability to the subject matter.

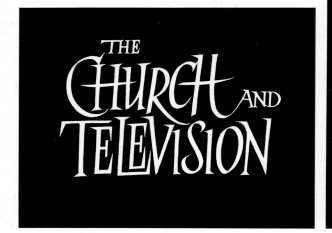

A single Masseeley caption and, below, part of a hand-set roller caption.

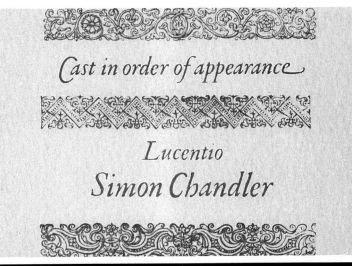

Cast in order of appearance

Lucentio
Simon Chandler

Commentary
TONY BASTABLE
ALAN GORE

Director
RICHARD MERVYN

PRODUCER'S
ASSISTANT
Cherry Britton

RESEARCH
OVEN
PETER JAMES
with
tomato s...
80
HORIZON

PAT
1979
3081

Photograph/Aston Electronic Designs

1

2

Electronic character generators

The very earliest character generators could not reproduce recognized fount styles; they could produce lettering only in low-resolution shapes, convenient to the technology available . . . not surprisingly, television graphic designers used these devices as infrequently as possible. *John H. Wood and Donald R. MacClymont SMPTE Technical Conference/Los Angeles/1985*

That has changed. Advances in micro-processing and components with vast memories have given improved definition, removed all positional and spacing constraints and finally given anti-aliasing. An operator can now, as shown left, **(1)**, select a typeface from a library, usually stored on a floppy disk. The lettering required is composed, in the size and style required on a keyboard similar to a typewriter and it appears *immediately* on a monitor. When the letterspacing and wordspacing and layout are exactly as required the wholeframe is stored in the computer's memory for instant re-call for transmission or editing.

These devices are now widespread and the convenience of keyboarding, at typing speed, bottom-of-frame information in almost any typeface only seconds before they can be revealed on-screen — especially for sports results and news programmes — has made the old handset type almost obsolete. Printing is now only used for special effects on artwork, printing on to cels for filming, or in emergencies. The Aston IV **(2)** has a keyboard, computer and disc drive and this was the first commercially available character generator designed to use anti-aliasing. It provides three levels in the character composition mode so that a single frame may have a background design, a second level with fixed displayed message, as the example **(3)** and a third level could be a moving roller giving more copy. The off-screen shot **(4)**, displaying a complete fount, one of over a thousand faces in the Aston IV library. It shows the fine serifs and thin strokes of an alphabet like Windsor Light can be accurately presented. Standard types are stored at 24, 32, 40 and 48 tv-lines.

Adding founts to the generator

Electronic character generators have given graphic designers the freedom to produce their own on-screen lettering very quickly and economically. A whole fount can be produced in a matter of a few hours. Even with the offer of very wide ranges of stock typefaces there are many occasions when special letterforms can enhance a design. Both the Aston III and IV have Font Compose systems where typefaces designed for, or by, the graphic designer can be digitized into the generator's memory and stored on to a floppy disk for later use.

The illustrations show: (5) a video camera input with artwork lettering and (6) a letter 'E' on the compose monitor at 32 tv-lines and the cursor by the middle horizontal. The operator controls the cursor to add, or remove pixels to refine the shape of the letter. Aston III was used to prepare the special face (7) designed by Mick Mannveille of Thames Television for a programme on George Orwell entitled '1984'.

5

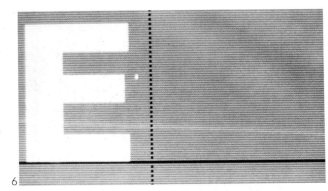

6

3

Windsor Light
ABCDEFGHIJKLMN
OPQRSTUVWXYZ
abcdefghijklmn
opqrstuvwxyz
1234567890
(!£$%&?/*:˜¿øØˆß)

4

7

8

Anti-aliased characters

The process used to conceal the very jagged edges caused by the size of the pixels in digitally generated letters is called 'anti-aliasing'. The generator is programmed to imitate the quality of the softened edges inherent in the camera's reproduction of artwork.

A magnified image of a 42 tv-line letter (8) where the upper half shows the digitized 'jaggies' and the lower half the anti-aliased effect plus its appearance on-screen produced on the Aston IV. Graphic designers in television now have character generators offering the facility and freedom developed by phototypesetting machines in printing.

Just as all television programmes have on-screen credits for those who take part and produce them – here are the credits for this book.

I have been helped by a large number of people throughout the world and I am pleased to acknowledge my gratitude to them.

To all the graphic designers, animators, programers and technicians whose work is represented; also to those who were generous enough to demonstrate that graphic designers can *write* and give reality and richness by their efforts, my sincere thanks for their contributions to 'Graphic Descriptions'.

To those in broadcast television my thanks to all the following for their contributions and advice: to Patrick Downing, Controller of Visual Services at Thames Television, who encouraged me to start this project and backed this up with many facilities throughout the long process of collating and writing: Brian Trigidden, Head of Graphic Design/BBC and his collegue John Aston, Design Manager, whose enthusiasm traced early filmed graphic work and much other valuable material: Tony Oldfield of LWT who charmed his designers to contribute both words and pictures, as did John Dee of Yorkshire: Sue Dix, Graphic Design Manager of The BBC Open University Production Centre: John Leech of Granada TV: Francis Heaney, Art Director of Headline News/Atlanta: Lou Dorfsman of

CBS/New York: and Ben Blank, Director of Graphics ABC News/New York. I received prompt help from Alan Hondow, Supervisor of Graphics and Kathy Day, Art Director of News Graphics, both of the Australian Broadcasting Corporation in Sydney. Dr John Emett and Vernon Smith of Thames were kind enough to give me technical and historical information. From Hans Donner of TV Globo I received many stills and a Telex message of 'novel' length; Gerard Bueche Art Director of the Canadian Broadcasting Corporation in Toronto: Ethan Ames, Head of Graphic Design at TVam: to David James of HTV for his care and interest: Peter Atkinson of Independent Television News for a very useful interview: Marc Ortmans and Simon Broom of Channel 4 for good discussions and much material: and to Michael Graham-Smith of the BBC for encouragement and listening.

From the independent design and production companies my thanks go to: Martin Lambie-Nairn who talked and provided stills; Bernard Lodge who did the same: Saul Bass who spoke and wrote so willingly: Darrell Pockett and his partners at English, Markell Pockett: Odile Santos of Parigraph for useful information; Catherine Murphy O'Connor and Robert van der Leeden of Ampex: Felicity Chadwick of · Crown Computer Graphics: Peter Florence of Digital Pictures: Roger Thornton and Richard Bain of Quantel: Veronique Damien of Sogitec/France: Jilly Knight of

Electronic Arts: Peter Stothart, General Manager at Cal Videographics: Paul Docherty of Electric Image: Graeme Scott and John Wakeford, Sales Director of Aston Electronic Designs: Shari Folz of Pacific Data Images: Giovanni della Rossa of Eidos/Milan: From California I received material from John C. Le Prevost: Tony Haines of Filmfex gave his time generously: Peter Truckel of The Moving Picture Company gave valuable information on computer-controlled model animation:

Janice, my wife, transferred my many drafts via a computer and high-speed printer and still wondered why it took so long. Roger Read, Senior Photographic Technician at Thames and his team were very helpful, as have they have been over the many years I have worked with them. My editor at Trefoil Books was John Latimer Smith and his skill, patience, and good company at luncheon enabled me to keep going.

Lastly, my thanks to friends who have been involved in ways which can not be detailed, and to my colleagues in television graphics – designers, technicians, photographers, rostrum operators and animators – from whom I have tried to learn.

Gathering reproduceable material from film and video sources from so many stations proved very difficult. Any errors in of fact and attribution are due entirely to me.

Douglas Merritt Teddington/June 1987

Further reading and books consulted

Expanded Cinema	*Gene Youngblood*	Studio Vista/1970
TV Graphics	*Roy Laughton*	Studio Vista/Van Nostrand Reinhold 1966
Film and TV Graphics	*Walter Herdeg and John Halas*	Graphis Press/1967
The Changing Image	*Geoffrey Crook*	Robots Press/1986
Creative Computer Graphics	*Annabel Jankel and Rocky Morton*	Cambridge University Press/1984
Dictionary of Computer Graphics	*John Vince*	Frances Pinter/1984
Television – the first fifty years	*Jeff Greenfield*	Harry N Abrams Inc/New York/1977
The Technique of Film Animation	*John Halas and Roger Manvell*	Focal Press/1979